CRM Short Courses

The volumes in the **CRM Short Courses** series have a primarily instructional aim, focusing on presenting topics of current interest to readers ranging from graduate students to experienced researchers in the mathematical sciences. Each text is aimed at bringing the reader to the forefront of research in a particular area or field, and can consist of one or several courses with a unified theme. The inclusion of exercises, while welcome, is not strictly required. Publications are largely but not exclusively, based on schools, instructional workshops and lecture series hosted by, or affiliated with, the *Centre de Researches Mathématiques* (CRM). Special emphasis is given to the quality of exposition and pedagogical value of each text.

More information about this series at http://www.springer.com/series/15360

Yuri I. Manin

Quantum Groups and Noncommutative Geometry

Second Edition

With a Contribution by Theo Raedschelders
and Michel Van den Bergh

Yuri I. Manin
Max Planck Institute for Mathematics
Bonn, Germany

ISSN 2522-5200 ISSN 2522-5219 (electronic)
CRM Short Courses
ISBN 978-3-030-07432-6 ISBN 978-3-319-97987-8 (eBook)
https://doi.org/10.1007/978-3-319-97987-8

Mathematics Subject Classification (2010): 16S10, 16S37, 16S38, 16T05, 16T15, 16W30, 18D50, 18D10, 20C30, 20G42

This Springer imprint is published by the registered company Springer Nature Switzerland AG
The registered company address is: Gewerbestrasse 11, 6330 Cham, Switzerland

Preface to the Second Edition

This second edition was prepared with great help of Th. Raedschelders, M. Van den Bergh, and D. Leites.

Th. Raedschelders and M. Van den Bergh carefully read the whole text and suggested many editorial corrections. D. Leites, who started working on this book when its Russian translation was being prepared about two decades ago, returned to this project now and did a very meticulous job.

I am deeply grateful to them.

For some mathematical updates, see Chapter 1 "Introduction".

Bonn, Germany

Yuri I. Manin

Contents

Chapter 1
Introduction

We begin with some terminology and background, in particular we define the notions of Hopf algebras and quantum groups.

Let H be the algebra of functions of a Lie group G. Then the multiplication map $G \times G \to G$ (resp. the inversion map $G \to G : g \to g^{-1}$, resp. the inclusion of the identity point) gives rise to a "comultiplication" $\Delta : H \to H \otimes H$ (resp. "antipode" $i : H \to H$, resp. "counit" $\varepsilon : H \to \mathbb{R}$), which satisfies a set of identities defining the general algebraic notion of a Hopf algebra (cf. Chapter 3 below). This set is symmetric with respect to reversing all arrows. However, an asymmetry might arise: while H, as algebra, always has a commutative multiplication, it may well have a noncocommutative comultiplication (this is the case if G is nonabelian).

Hopf algebras that are not necessarily commutative or cocommutative have been studied by algebraists for several decades (cf. Abe [1]). Recently, however, some very specific Hopf algebras emerged in mathematical physics; V.G. Drinfeld named them "quantum groups." Initially, they appeared in the quantum inverse scattering transform method (QIST) developed by L.D. Faddeev and his school (cf. [26, 27, 56, 58, 59]) and reviewed in V.G. Drinfeld's Berkeley talk [21]. Closely related work (also motivated by QIST) was done by M. Jimbo [36, 37] and, from a somewhat different viewpoint, by S.L. Woronowicz [68, 69]. One of the main ideas behind these works is that such rigid objects as classical simple groups (or Lie algebras) admit in fact continuous deformations at the level of Hopf algebras corresponding to them, and the deformed objects close to the initial one can be described very precisely together with their representation theory.

In this work, we systematically develop a different approach to quantum groups, based upon the following observation. Suppose that you "quantize" the simplest phase plane, imposing on its coordinates the commutation relation

$$xy = e^{\hbar} yx ;$$

Y. I. Manin, *Quantum Groups and Noncommutative Geometry*,
CRM Short Courses, https://doi.org/10.1007/978-3-319-97987-8_1

i.e., the *integrated* version of the Heisenberg commutation relation. Then the ordinary symmetry group GL(2) of the plane breaks down. However the "broken symmetry" is completely restored if one imposes some nontrivial commutation relations upon the entries of the (2×2)-matrices, the elements of GL(2). In this way, one arrives at the notion of the quantum group $\mathrm{GL}_q(2)$, where $q = e^{\hbar}$, which is described in complete detail in Chapter 2.

One remarkable property of this approach is its generality. Namely, in Chapters 4 and 5, instead of "quantum plane," we start with a "quantum linear space" defined by **arbitrary quadratic relations** between its noncommutative coordinates. In this way we obtain a "general linear quantum group," or rather a pair:

"quantum semigroup of endomorphisms" $\rightarrow$ "quantum group."

The first object is a noncommutative space of the same kind (i.e., it is defined by quadratic relations) while the second one is obtained from the first one by a process of noncommutative localization as advocated in algebra by P.M. Cohn, and arising here for the first time in a natural way. The point is that one inverts matrices and not just elements of a ring, and to obtain a Hopf algebra, one must, generally speaking, invert infinitely many matrices.

We use the word "noncommutative space" in the spirit of Alain Connes. The main difference is that we develop a fragment of noncommutative *algebraic* geometry while Connes deals with noncommutative *differential* geometry and *topology*. In Chapter 11, we discuss a way to introduce a $*$-structure in our groups thus making it possible to define their compact forms.

As in [56, 68, 69], but unlike [21], we work with *the noncommutative ring of functions over a quantum group* rather than with the "universal enveloping algebra," which is a dual object. Of course, both objects deserve close attention, but in our approach the former appears more naturally. The key technical notion in this connection is that of a multiplicative matrix: cf. Sections 3.6–3.10. It is also worth mentioning that we have no need to consider only small deformations of classical objects: the parameter spaces of our objects are defined globally. They are just Grassmannians of quadratic relations: cf. Section 4.2. The price paid for this is the loss of the notion of a "semisimple" quantum group (which anyway has never been formalized in previous works).

Moreover, we do not need to impose any relation on the Yang–Baxter equations from the start. However, we can and must explain such a relation at a later stage. An important thing to remember is that a Yang–Baxter operator generates quantum groups by two very different procedures. One is to consider the quantum automorphism group of a quadratic algebra which is just a "Yang–Baxter symmetric algebra." The groups appearing in QIST are of this type.

Another way starts with a relativization of the notions of quadratic algebra, quantum group, etc. by replacing everywhere the transposition of factors in a tensor product by a Yang–Baxter operator. The simplest example is the "super version" of our construction, which is fairly obvious. Other examples were not considered before [46]. We explain these ideas rather succinctly in Chapters 12 and 13.

An earlier version of this work is [46]. Many details and some new results are added here. I would like to mention in particular the general construction of a Hopf algebra starting from a bialgebra with a generating multiplicative matrix (see Chapter 8). These notes are not meant to be a survey of this quickly growing subject; the bibliography is very incomplete.[1] The ideas of this paper were first developed in my lectures in Moscow University during the winter of 1986–1987 and the subsequent seminar where Yu. Kobyzev suggested the description of $\mathrm{GL}_q(2)$ which was seminal for all that follows.

These notes were written for a series of lectures given at the Centre de recherches mathématiques at the Université de Montréal in June 1988, while I was invited as Aisenstadt Professor. I am grateful to many people who made my stay in Montréal very agreeable and productive, in particular to André Aisenstadt, whom it was my pleasure to meet. I would like to thank Luis Alvárez-Gaumé and John Harnad for their assistance in the preparation of these notes. Finally I would like to thank Nathalie Brunet, Louise Letendre, and Angèle Patenaude for their careful typing of the manuscript.

[1] Several text books and reviews on quantum groups were published: [14, 19, 39, 40, 45]. They contain a more complete bibliography and answer some of the questions posed in Chapter 14 of this book.

Chapter 2
The Quantum Group $\mathrm{GL}_q(2)$

2.1 Noncommutative Spaces: Points and Rings of Functions

In this paper, we fix once and for all a field $\mathbb{K}$. A *ring* (or an *algebra*) means an associative $\mathbb{K}$-algebra with unit, not necessarily commutative. It is suggestive to imagine the ring A as a ring of (polynomial) functions on a space which is an object of noncommutative, or "quantum," geometry. Morphisms of spaces correspond to ring homomorphisms in the opposite direction. For A and B fixed, the set $\mathrm{Hom}_{\mathbb{K}-\mathrm{alg}}(A, B)$ is also called the set of *B-points of the space* defined by A.

There is no harm in giving a formal definition of the category of noncommutative spaces as $(\mathbb{K}-\mathrm{Alg})^{\mathrm{op}}$ as long as one remembers that some of the most common categorical prejudices could be misleading when working with $(\mathbb{K}-\mathrm{Alg})^{\mathrm{op}}$. To quote just one, the tensor product in $\mathbb{K}-\mathrm{Alg}$ *does not* define a direct product in $(\mathbb{K}-\mathrm{Alg})^{\mathrm{op}}$, but morally *does correspond to a "direct product"* of quantum spaces.

2.2 Two Quantum Planes and Quantum Matrices

Fix $q \in \mathbb{K}$, where $q \neq 0$. The *quantum plane* $A_q^{2|0}$ is defined by the ring

$$A_q^{2|0} := \mathbb{K}\langle x, y\rangle/(xy - q^{-1}yx) , \tag{2.1}$$

where $\mathbb{K}\langle x_1, \dots, x_n\rangle$ is the associative $\mathbb{K}$-algebra freely generated by $x_1, \dots, x_n$. This plane $A_q^{2|0}$ is a deformation of the usual plane corresponding to $q = 1$. We also need a deformation of the $0|2$-dimensional "plane" of supergeometry:

$$A_q^{0|2} := \mathbb{K}\langle \xi, \eta\rangle/(\xi^2, \eta^2, \xi\eta + q\eta\xi) . \tag{2.2}$$

Y. I. Manin, *Quantum Groups and Noncommutative Geometry*,
CRM Short Courses, https://doi.org/10.1007/978-3-319-97987-8_2

For $q = 1$, this superplane is the Grassmann algebra with 2 generators. Both rings (2.1) and (2.2) are naturally graded (the generators being of degree 1), and the dimensions of their homogeneous components coincide with those of the symmetric and exterior algebra, respectively. In fact, the monomials $x^a y^b$, where $0 \leq a, b < \infty$ (resp. the monomials $\xi^a \eta^b$, where $0 \leq a, b \leq 1$) form a $\mathbb{K}$-basis of (2.1) (resp. of (2.2)).

Finally, the coordinate ring of the manifold of quantum (2×2)-matrices $\left(\begin{smallmatrix} a & b \\ c & d \end{smallmatrix}\right)$ is defined by $M_q(2) := \mathbb{K}\langle a, b, c, d \rangle / I$, where

$$I = \big(ab - q^{-1}ba, ac - q^{-1}ca, cd - q^{-1}dc, bd - q^{-1}db, \\ bc - cb, ad - da - (q^{-1} - q)bc\big) . \quad (2.3)$$

Although the commutation relations (2.3) are slightly more complicated than (2.1), one easily proves that the monomials $a^\alpha b^\beta c^\gamma d^\delta$ still form a basis of $M_q(2)$.

In order to state our main theorem on $M_q(2)$, we need the following notation. Two families S and T of elements of a ring A are called *commuting* if $[s, t] = 0$ for any $s \in S, t \in T$. An A-point of $A_q^{2|0}$, $A_q^{0|2}$, $M_q(2)$, etc., is considered as a family of its coordinates, e.g., an A-point of $M_q(2)$ is a quadruple $(a, b, c, d) \in A^4$ satisfying (2.3). In this sense, we can say that two points are commuting. Finally, a B-point of a quantum space A is called a *generic point*, if the corresponding morphism $A \to B$ is an injection.

2.3 Theorem

(a) *Let* $\left(\begin{smallmatrix} a & b \\ c & d \end{smallmatrix}\right), \left(\begin{smallmatrix} a' & b' \\ c' & d' \end{smallmatrix}\right)$ *be two commuting A-points of* $M_q(2)$. *Then* $\left(\begin{smallmatrix} a & b \\ c & d \end{smallmatrix}\right)\left(\begin{smallmatrix} a' & b' \\ c' & d' \end{smallmatrix}\right)$ *is an A-point of* $M_q(2)$.

(b) *With the same assumption, let*

$$\mathrm{DET}_q \begin{pmatrix} a & b \\ c & d \end{pmatrix} := ad - q^{-1}bc = da - qcb .$$

Then

$$\mathrm{DET}_q \left[\begin{pmatrix} a & b \\ c & d \end{pmatrix} \begin{pmatrix} a' & b' \\ c' & d' \end{pmatrix} \right] = \mathrm{DET}_q \begin{pmatrix} a & b \\ c & d \end{pmatrix} \mathrm{DET}_q \begin{pmatrix} a' & b' \\ c' & d' \end{pmatrix} , \quad (2.4)$$

and $\mathrm{DET}_q\left(\begin{smallmatrix} a & b \\ c & d \end{smallmatrix}\right)$ *commutes with* a, b, c, d.

(c) *Assume in addition that* $\mathrm{DET}_q\left(\begin{smallmatrix} a & b \\ c & d \end{smallmatrix}\right)$ *is invertible in A. Then*

$$\begin{pmatrix} a & b \\ c & d \end{pmatrix}^{-1} = \left(\mathrm{DET}_q \begin{pmatrix} a & b \\ c & d \end{pmatrix}\right)^{-1} \begin{pmatrix} d & -qb \\ -q^{-1}c & a \end{pmatrix} \tag{2.5}$$

is an A-point of $M_{q^{-1}}(2)$.

In principle, all these statements can be checked directly; e.g., the last statement is quite straightforward. However, checking the commutation relations (2.3) for the product of two commuting points is cumbersome and not very illuminating. One does not readily see what is so special about (2.3).

A proper way to understand (2.3) was suggested by Yu. Kobyzev. After all, matrices act on coordinate spaces, and matrix multiplication is just a reflection of this action. The same happens, actually, in the noncommutative realm.

2.4 Theorem

Let (x, y) *(resp.* (ξ, η)*) be a generic A-point of* $A_q^{2|0}$ *(resp.* $A_q^{0|2}$*) and suppose* $(a, b, c, d) \in A^4$ *commutes with* (x, y, ξ, η)*. Write*

$$\begin{pmatrix} x' \\ y' \end{pmatrix} = \begin{pmatrix} a & b \\ c & d \end{pmatrix}\begin{pmatrix} x \\ y \end{pmatrix}, \quad \begin{pmatrix} x'' \\ y'' \end{pmatrix} = \begin{pmatrix} a & c \\ b & d \end{pmatrix}\begin{pmatrix} x \\ y \end{pmatrix}, \quad \begin{pmatrix} \xi' \\ \eta' \end{pmatrix} = \begin{pmatrix} a & b \\ c & d \end{pmatrix}\begin{pmatrix} \xi \\ \eta \end{pmatrix}.$$

If $q^2 \neq -1$, *the following conditions are equivalent:*

(i) (x', y') *and* (x'', y'') *are points of* $A_q^{2|0}$;
(ii) (x', y') *is a point of* $A_q^{2|0}$, *while* (ξ', η') *is a point of* $A_q^{0|2}$;
(iii) (a, b, c, d) *is a point of* $M_q(2)$.

For $q^2 = -1$, *we have only* (iii) $\Longrightarrow$ (i) *and* (iii) $\Longrightarrow$ (ii).

Proof. Let us check, e.g., that (i) $\Longleftrightarrow$ (iii). The relation $x'y' = q^{-1}y'x'$ means

$$(ax + by)(cx + dy) = q^{-1}(cx + dy)(ax + by) . \tag{2.6}$$

Since (x, y) is a generic point and a, b, c, d commute with x, y, the equality (2.6) is equivalent to the equality of the three coefficients:

$$\begin{aligned} x^2 &: ac = q^{-1}ca , \\ y^2 &: bd = q^{-1}db , \\ xy &: ad - da = q^{-1}cb - qbc . \end{aligned} \tag{2.7}$$

Interchanging b and c, we get the relations equivalent to $x''y'' = q^{-1}y''x''$:

$$ab = q^{-1}ba\ , \qquad cd = q^{-1}dc\ , \qquad ad - da = q^{-1}bc - qcb\ . \tag{2.8}$$

Comparing the last relations in (2.7) and (2.8), we obtain

$$(q + q^{-1})(bc - cb) = 0 \Longrightarrow bc = cb\ , \quad \text{if } q^2 \neq -1\ .$$

Hence (2.7) and (2.8) together are equivalent to (2.3), if $q^2 \neq -1$.

A similar direct calculation shows that

$$(\xi', \eta') \text{ is a point of } A_q^{0|2} \Longleftrightarrow (2.8)$$

thus proving (i) $\Longleftrightarrow$ (ii). □

2.5 Proof of Theorem 2.3

We can now prove the multiplicativity property in a natural way. Take a generic point (x, y) of $A_q^{2|0}$ in a ring containing (a, b, c, d) and (a', b', c', d'). We can find such a ring and a point, commuting with (a, b, c, d) and (a', b', c', d'), e.g., $\mathbb{K}[a, \ldots, d'] \otimes A_q^{2|0}$ will do. Then $\left(\begin{smallmatrix} a & b \\ c & d \end{smallmatrix}\right)\left(\begin{smallmatrix} x \\ y \end{smallmatrix}\right)$ is a point of $M_q(2)$ commuting with (a', b', c', d'). It is also generic since it becomes generic after a specialization $\left(\begin{smallmatrix} a & b \\ c & d \end{smallmatrix}\right) \rightarrow \left(\begin{smallmatrix} 1 & 0 \\ 0 & 1 \end{smallmatrix}\right)$. Therefore $\left(\begin{smallmatrix} a' & b' \\ c' & d' \end{smallmatrix}\right)\left(\begin{smallmatrix} a & b \\ c & d \end{smallmatrix}\right)\left(\begin{smallmatrix} x \\ y \end{smallmatrix}\right)$ is a point of $M_q(2)$. Hence $\left(\begin{smallmatrix} a' & b' \\ c' & d' \end{smallmatrix}\right)\left(\begin{smallmatrix} a & b \\ c & d \end{smallmatrix}\right)$ satisfies (2.7). One similarly proves (2.8), using $A_q^{0|2}$. This reasoning is valid only for $q^2 \neq -1$, but the result is also true for $q^2 = -1$ in view of the *extension principle of algebraic identities.*

We also get a natural definition of the quantum determinant which proves its multiplicativity: in the notation of Theorem 2.4,

$$\xi'\eta' = (a\xi + b\eta)(c\xi + d\eta) = \mathrm{DET}_q \begin{pmatrix} a & b \\ c & d \end{pmatrix} \xi\eta\ .$$

□

2.6 Quantum Analogues of Groups GL(2) and SL(2) and Their Representations

We can now define spaces of functions on quantum groups imitating the classical procedure of inverting DET_q or equating it to 1. (For a justification and a more sophisticated treatment, see Chapters 3 and 8.) We set

$$\mathrm{GL}_q(2) : M_q(2)[t]/([t,a],[t,b],[t,c],[t,d]; t\,\mathrm{DET}_q - 1)\,,$$
$$\mathrm{SL}_q(2) : M_q(2)/(\mathrm{DET}_q - 1)\,.$$

Theorem 2.4 describes the representations of these quantum groups in quantum spaces $A_q^{2|0}$ and $A_q^{0|2}$.

There are two ways of looking at these representations. One can consider, say, the whole of $A_q^{2|0}$, the noncommutative space on which $\mathrm{GL}_q(2)$ acts, an analogue of the tautological two-dimensional representation.

Alternatively, one can consider only its linear part $\left(A_q^{2|0}\right)_1 = \mathbb{K}x \oplus \mathbb{K}y$ as the analogue. Then $\left(A_q^{2|0}\right)_d$ corresponds to the "quantum dth symmetric power" of the tautological representation and A to a kind of bosonic Fock space.

Actually, these two viewpoints can and should be reconciled in the notion that the ring of functions on the quantum dth symmetric power of the tautological representation is the "Veronese subring" $\bigoplus_{i \geq 0} \left(A_q^{2|0}\right)_{di}$.

2.7 Further Developments

The rest of this book contains extensions of this example in various directions. We show that one can replace $A_q^{2|0}$ by an arbitrary quadratic algebra and still define a reasonable quantum semigroup acting on it. In order to turn this semigroup into a quantum group, we have to work a bit harder than in our example (or in the commutative case) since in general there is no determinant with the necessary properties.

The whole initial data can be generalized so as to include, e.g., quantum deformations of linear *supergroups*. Actually, the correct way to do this is to develop an axiomatic theory of exact tensor categories. In this way, we get a better understanding of the Yang–Baxter (or triangle) equations and their role in the construction of quantum groups.

Finally, we initiate the study of homological properties of quantum spaces and groups.

Chapter 3
Bialgebras and Hopf Algebras

3.1 Bialgebras

Let H be a $\mathbb{K}$-module. Recall that a *bialgebra* structure on H is defined by four morphisms

$$H \otimes H \xrightarrow{m} H \xrightarrow{\Delta} H \otimes H \,,$$
$$\mathbb{K} \xrightarrow{\eta} H \xrightarrow{\varepsilon} \mathbb{K} \,,$$

satisfying the following axioms, written as commutative diagrams:

Algebras
Associativity:

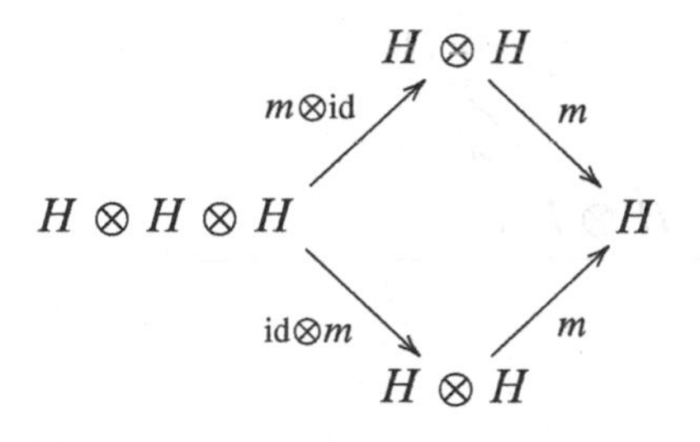

Coalgebras
Coassociativity:

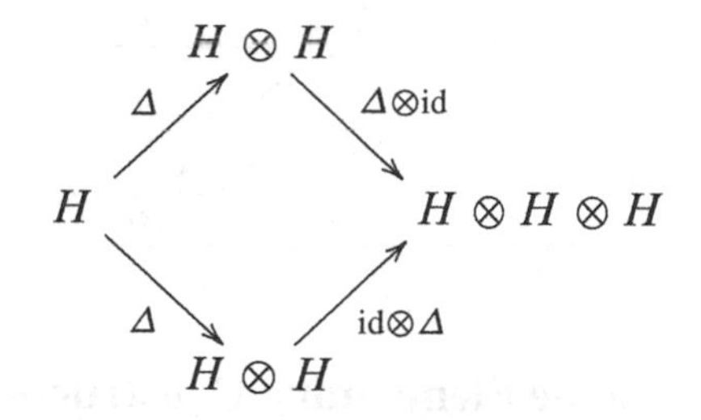

Unit:

$H \otimes H$

$\mathrm{id}\otimes\eta,\ \eta\otimes\mathrm{id}$ ↗ ↘ m

$$H = H \otimes \mathbb{K} = \mathbb{K} \otimes H \xrightarrow{\mathrm{id}} H$$

Counit:

$H \otimes H$

Δ ↗ ↘ $\varepsilon\otimes\mathrm{id},\ \mathrm{id}\otimes\varepsilon$

$$H \xrightarrow{\mathrm{id}} H = H \otimes \mathbb{K} = \mathbb{K} \otimes H$$

Y. I. Manin, *Quantum Groups and Noncommutative Geometry*,
CRM Short Courses, https://doi.org/10.1007/978-3-319-97987-8_3

Connecting axiom:

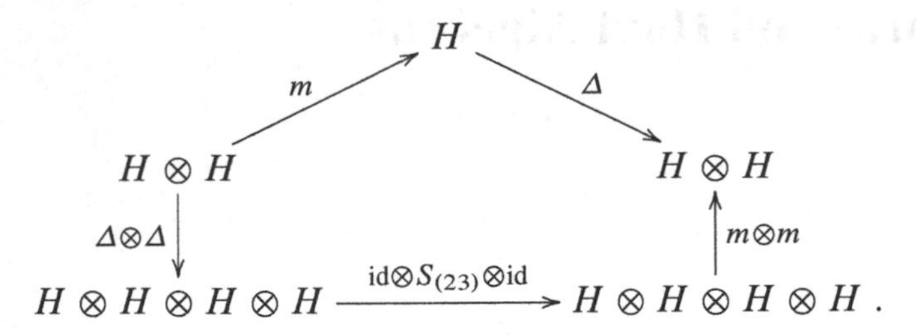

Here S_σ denotes the canonical morphism corresponding to a permutation σ. The connecting axiom means that Δ is an algebra morphism or, equivalently, that m is a coalgebra morphisms. Its diagrammatic form makes its self-dual nature evident.

Remark. When convenient, we will further assume that Δ is a unital algebra morphism, η a counital coalgebra morphism, and $\epsilon \circ \eta = \mathrm{id}$. This amounts to the commutativity of three more diagrams.

3.2 Antipode

An *antipode* of a bialgebra (H, m, Δ) is a linear map $i: H \to H$ such that the following diagram is commutative:

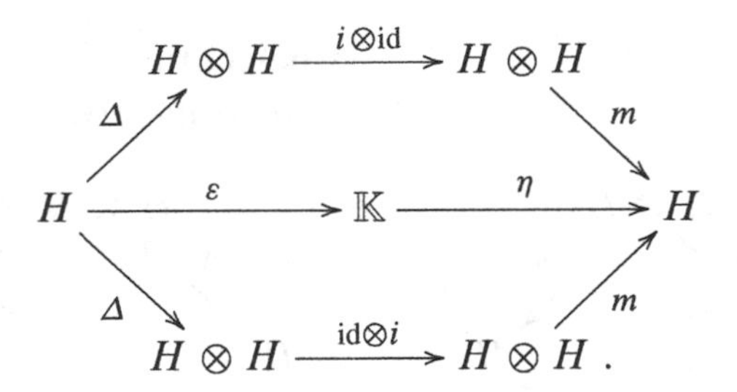

3.3 Some Elementary Constructions

Let

$$m^{\mathrm{op}} = m \circ S_{(12)}, \quad \Delta^{\mathrm{op}} = S_{(12)} \circ \Delta.$$

If (H, m, Δ) is a bialgebra, then $(H, m^{\mathrm{op}}, \Delta)$ and $(H, m, \Delta^{\mathrm{op}})$ are also bialgebras. If i is a bijective antipode for (H, m, Δ), then i^{-1} is a bijective antipode for $(H, m^{\mathrm{op}}, \Delta)$ and $(H, m, \Delta^{\mathrm{op}})$, hence i is a bijective antipode for $(H, m^{\mathrm{op}}, \Delta^{\mathrm{op}})$.

3.4 Theorem

If an antipode i exists, it is unique and reverses multiplication and comultiplication, i.e., it defines a bialgebra morphism (ε, η are not changed):

$$i\colon (H, m, \Delta) \to (H, m^{\mathrm{op}}, \Delta^{\mathrm{op}}).$$

In other words, we have commutative diagrams

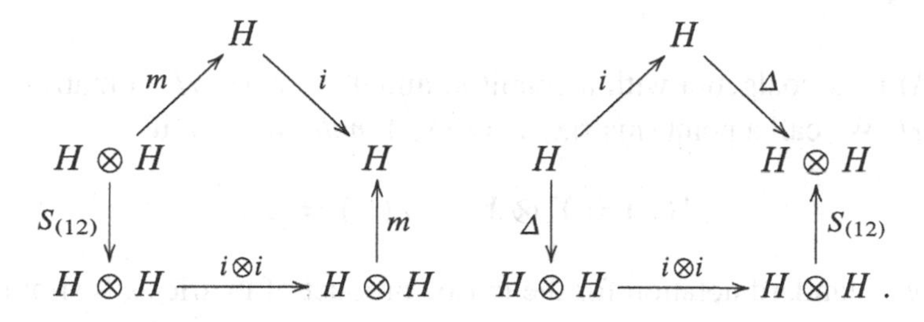

For a proof, cf. Abe [1].

3.5 Bialgebras and Quantum Groups

The ring of polynomial functions on an affine algebraic group G is a bialgebra with an antipode, the multiplication being induced by the group law $G \times G \to G$ and the antipode being induced by the inversion map $G \to G\colon x \mapsto x^{-1}$. This bialgebra is commutative. Dropping this condition of commutativity, we get the general notion of *Hopf algebra* which is a formalization of the (so far) intuitive notion of "quantum group."

The point functor $A \to \mathrm{Hom}_{\mathbb{K}-\mathrm{Alg}}(H, A)$ for any Hopf algebra (H, m, Δ, i) satisfies the same properties as those of $\mathrm{GL}_q(2)$ (cf. Theorem 2.3):

(a) Let $f, g\colon H \to A$ be two *commuting* points (i.e., $[f(h), g(h')] = 0$ for any $h, h' \in H$). Define their product as a composite map:

$$fg\colon H \xrightarrow{\Delta} H \otimes H \xrightarrow{f \otimes g} A \otimes A \xrightarrow{m} A\,.$$

(One needs commutativity to prove that fg is an algebra morphism.)

(b) Let $f\colon H \to A$ be a point. Then, if i is bijective,

$$f \circ i\colon H \xrightarrow{i} H \xrightarrow{f} A$$

is a point of $(H, m^{\mathrm{op}}, \Delta^{\mathrm{op}}, i)$. (Reversing multiplication and comultiplication maps for $\mathrm{GL}_q(2)$, we get $\mathrm{GL}_{q^{-1}}(2)$.)

Note that in a general bialgebra (i.e., in a *quantum semigroup*) i may not exist (as in $M_q(2)$); if it exists, it might be not bijective. If it is bijective, it may happen that $i^2 \neq \mathrm{id}$.

3.6 Multiplicative Matrices

Let (H, Δ) be a coalgebra with a counit ε, and $Y \in M(n, H)$ a matrix with elements in H. We call a point (matrix) $Y = (y_i^j)$ *multiplicative* if

$$\Delta(Y) = Y \otimes Y\,, \quad \varepsilon(Y) = E\,. \tag{3.1}$$

This is *not* a standard notation for the tensor product of matrices. It simply means that

$$\Delta\big(y_i^j\big) = \sum_k y_i^k \otimes y_k^j\,, \quad \varepsilon\big(y_i^j\big) = \delta_i^j\,.$$

Examples. Let $\left(\begin{smallmatrix} a & b \\ c & d \end{smallmatrix}\right) \in M_q(2)$ and $\left(\begin{smallmatrix} a & b & 0 \\ c & d & 0 \\ 0 & 0 & t \end{smallmatrix}\right) \in \mathrm{GL}_q(2)$. If a quantum group H admits a multiplicative matrix Y whose entries generate H as a ring, then H can be called a "quantum matrix group" (cf. Woronowicz [68, 69], who uses a special sign instead of our fake tensor product). Below we give a representation-theoretical interpretation of multiplicative matrices, but first we state some of their properties, which will help us in Chapter 8 to construct antipodal maps.

3.7 Proposition

(a) $\Delta(Y) = Y \otimes Y \iff \Delta^{\mathrm{op}}(Y^{\mathrm{t}}) = Y^{\mathrm{t}} \otimes Y^{\mathrm{t}}$.

(b) *Assume that* Y *is a multiplicative matrix in a Hopf algebra with antipode* i. *Define* $Y_k = i^k(Y)$. *Then*

$$\begin{aligned} Y_k Y_{k+1} &= Y_{k+1} Y_k = E \qquad \textit{for } k \equiv 0 \pmod 2\,,\\ Y_k^{\mathrm{t}} Y_{k+1}^{\mathrm{t}} &= Y_{k+1}^{\mathrm{t}} Y_k^{\mathrm{t}} = E \qquad \textit{for } k \equiv 1 \pmod 2\,, \end{aligned} \tag{3.2}$$

and

$$\Delta(Y_k) = \begin{cases} Y_k \otimes Y_k & \textit{for } k \equiv 0 \pmod 2 \\ (Y_k^{\mathrm{t}} \otimes Y_k^{\mathrm{t}})^{\mathrm{t}} & \textit{for } k \equiv 1 \pmod 2. \end{cases} \tag{3.3}$$

(c) *If f, g are commuting A-points of H, as in Section* 3.5, *then*

$$(fg)(Y) = f(Y)g(Y) .$$

Proof. (a) Let $\Delta(Y) = Y \otimes Y$. Then

$$\begin{aligned}
\big(\Delta^{\mathrm{op}}(Y^{\mathrm{t}})\big)_i^k &= \Delta^{\mathrm{op}}(Y)_k^i = S_{(12)} \circ \Delta(Y)_k^i \\
&= S_{(12)}\Big(\sum_j y_k^j \otimes y_j^i\Big) = \sum_j y_j^i \otimes y_k^j ; \\
(Y^{\mathrm{t}} \otimes Y^{\mathrm{t}})_i^k &= \sum_j (Y^{\mathrm{t}})_i^j (Y^{\mathrm{t}})_j^k = \sum_j y_j^i \otimes y_k^j .
\end{aligned}$$

(b) Applying the antipode axiom (Section 3.2) to Y, we get

$$i(Y)Y = Yi(Y) = I .$$

Since i reverses the multiplication map, we have $i(AB) = \big(i(B^{\mathrm{t}})y(A^{\mathrm{t}})\big)^{\mathrm{t}}$ for any two matrices $A, B \in H$. From this we obtain (3.2) by induction. Finally, again by induction

$$\begin{aligned}
\Delta(Y_{k+1}) &= \Delta \circ i(Y_k) = S_{(12)}(i \otimes i)\big(\Delta(Y_k)\big) \\
&= \begin{cases} S_{(12)}(Y_{k+1} \otimes Y_{k+1}) & \text{for } k \equiv 0 \pmod 2 \\ S_{(12)}\big((Y_{k+1}^{\mathrm{t}} \otimes Y_{k+1}^{\mathrm{t}})\big)^{\mathrm{t}} & \text{for } k \equiv 1 \pmod 2 . \end{cases}
\end{aligned}$$

This establishes (3.3) since the computation at the beginning of our proof shows that $S_{(12)}(Y_k \otimes Y_k) = (Y_k^{\mathrm{t}} \otimes Y_k^{\mathrm{t}})^{\mathrm{t}}$.

(c) By definition, and due to condition (3.1)

$$(fg)(Y) = m \circ (f \otimes g) \circ \Delta(Y) = f(Y)g(Y) .$$

□

3.8 Comodules

A *left comodule over a coalgebra* (H, Δ, ε) is a linear space M together with a morphism $\delta\colon M \to H \otimes M$, called *coaction*, such that the following two diagrams are commutative:

Coassociativity:

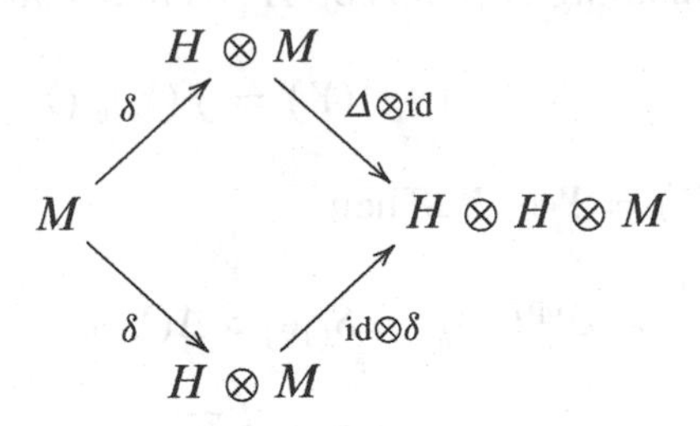

Counit:

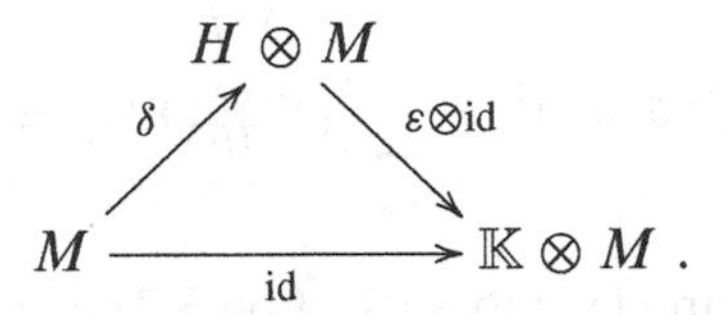

One similarly defines a *right comodule* by means of $\delta\colon M \to M \otimes H$.

Let (M, δ) be a left (H, Δ, ε)-comodule. Then $(M, \delta^{\mathrm{op}} = S_{(12)} \circ \delta)$ is a right $(H, \Delta^{\mathrm{op}}, \varepsilon)$-comodule, and vice versa. A morphism of comodules is defined in a natural way.

Let $(\mathbb{K}^n, \delta)$ be a finite-dimensional left H-comodule. Define a matrix Y by setting

$$\delta(e_i) = \sum_{j=1}^{n} y_i^j \otimes e_j \ , \tag{3.4}$$

where $\{e_j\}_{j=1}^n$ is a canonical basis of $\mathbb{K}^n$.

3.9 Proposition

(a) *This construction* (3.4) *establishes a bijection between all structures of a left comodule on* $\mathbb{K}^n$ *and multiplicative matrices in* $M(n, H)$.

(b) *A linear map* $f : (\mathbb{K}^n, \delta) \to (\mathbb{K}^m, \delta')$ *is a morphism of comodules given by the multiplicative matrices* Y *and* Y'*,if and only if*

$$FY' = YF \ , \quad \textit{where } F = \left(f_i^j\right) \textit{ and } f(e_i) = \sum_j f_i^j e'_j \ .$$

Proof. (a) Applying to $(\mathbb{K}^n, \delta)$ the coassociativity axiom, we obtain

$$\Delta(Y) = Y \otimes Y \ .$$

Similarly, applying the counit axiom, we get $\varepsilon(Y) = Y$. Clearly, the other way round, these conditions imply the axioms. The statement (b) is straightforward. □

3.10 Representations

In general, there are at least three versions of the notion of a "representation" of a quantum (semi)group H.

(a) *A left (or right) H-comodule.*

This corresponds to the usual notion of group representation if one looks at H as at the ring of functions on our group. Further on, we adopt *this* viewpoint.

(b) *A left (or right) H-module.*

One rarely considers this notion in the classical situation since (H, m) is then commutative. However, the classical representations define modules over the universal enveloping algebras dual to rings of functions. Thus, morally, H-modules are given by representations of a "dual quantum group."

(c) *A morphism of Hopf algebras $H' \to H$ (or, dually $H \to H'$).*

This viewpoint corresponds to the classical notion of, say, a unitary representation considered as a morphism $G \to \mathrm{U}(n)$.

Below, we construct quantum endomorphism semigroups (resp. automorphism groups) of general quadratic algebras which will provide us with representations in both senses (a) and (c), but not (b).

3.11 Generators and Relations

Algebras are often defined by generators and relations, as, say, $T(V)/I$, where V is a linear space, $T(V)$ its tensor algebra and I the ideal of relations.

A linear subspace $I \subset H$ is an *ideal* in (H, m) if

$$m(I \otimes H + H \otimes I) \subset I\ .$$

Dually, I is a *coideal* in the coalgebra (H, Δ) if

$$\Delta(I) \subset I \otimes H + H \otimes I\ .$$

A subspace I in a bialgebra which is both an ideal and a coideal induces a bialgebra structure on H/I. An antipode i_H survives in H/I if $i_H(I) \subset I$.

Chapter 4
Quadratic Algebras as Quantum Linear Spaces

4.1 Notation

A *quadratic algebra* is an associative graded algebra $A = \bigoplus_{i=0}^{\infty} A_i$ with the following properties:

- $A_0 = \mathbb{K}$ (the ground field);
- A is generated by A_1;
- the ideal of relations between elements of A_1 is generated by the subspace of all quadratic relations $R(A) \subset A_1^{\otimes 2}$.

It is convenient to write $A \leftrightarrow \{A_1, R(A)\}$. We assume $\dim A_1 < \infty$.

Quadratic algebras form a category QA: its morphisms $f : A \to B$ are in bijection with linear maps $f_1 : A_1 \to B_1$ such that

$$(f_1 \otimes f_1)\big(R(A)\big) \subset R(B)\ .$$

Therefore, we have a forgetful functor

$$\mathrm{QA} \to \mathbb{K}\text{–mod}\colon A \to A_1\ .$$

In the next section, we show that an arbitrary quadratic algebra can play the role of a "quantum plane" considered in Chapter 2.

4.2 Operations on Quadratic Algebras

For any $A, B \in \mathrm{Ob(QA)}$, define:

$$\tilde{A} \leftrightarrow \{A_1, \{0\}\}\ , \tag{4.1}$$

Y. I. Manin, *Quantum Groups and Noncommutative Geometry*,
CRM Short Courses, https://doi.org/10.1007/978-3-319-97987-8_4

$$A^{\mathrm{op}} \leftrightarrow \{A_1, S_{(12)}(R(A))\}\,, \tag{4.2}$$
$$A^! \leftrightarrow \{A_1^*, R(A)^\perp\}\,, \quad \text{more precisely, } \{\Pi(A_1^*), R(A)^\perp\}\,, \tag{4.3}$$
$$A \circ B \leftrightarrow \{A_1 \otimes B_1, S_{(23)}(R(A) \otimes B_1^{\otimes 2} + A_1^{\otimes 2} \otimes R(B))\}\,, \tag{4.4}$$
$$A \bullet B \leftrightarrow \{A_1 \otimes B_1, S_{(23)}(R(A) \otimes R(B))\}\,, \tag{4.5}$$
$$A \otimes B \leftrightarrow \{A_1 \oplus B_1, R(A) \oplus [A_1, B_1] \oplus R(B)\}\,, \tag{4.6}$$
$$A \underline{\otimes} B \leftrightarrow \{A_1 \oplus B_1, R(A) \oplus [A_1, B_1]_+ \oplus R(B)\}\,, \tag{4.7}$$

where in (4.6) (resp. (4.7)), the commutator $[A_1, B_1]$ (resp. the anticommutator $[A_1, B_1]_+$) denotes the subspace in $A_1 \otimes B_1 \oplus B_1 \otimes A_1$ generated by $a \otimes b - b \otimes a$ (resp. $a \otimes b + b \otimes a$).

Clearly, $\tilde{A}$ is the tensor algebra of the space A_1. The natural map $\tilde{A} \to A$, identical on A_1, identifies A with $\tilde{A}/R_A$, where $R_A = \bigoplus_{i=0}^{\infty} R_i(A)$, and where

$$\begin{aligned}
R_0(A) &:= \{0\}\,,\\
R_1(A) &:= \{0\}\,,\\
R_2(A) &:= R(A)\,,\\
R_n(A) &:= \sum_{i=0}^{n-2} A_1^{\otimes i} \otimes R(A) \otimes A_1^{\otimes(n-i-2)} \quad \text{for } n \geq 2\,.
\end{aligned} \tag{4.8}$$

In (4.3),[1] we define $A_1^* := \mathrm{Hom}(A_1, \mathbb{K})$, identify $(V \otimes W)^*$ with $V^* \otimes W^*$ by means of the formula

$$(f \otimes g)(a \otimes b) = f(a)g(b) \quad \text{for any } f \in V^*, g \in W^*, a \in V \text{ and } b \in W$$

and set

$$R(A)^\perp = \{q \in A_1^* \otimes A_1^* \mid q(r) = 0\} \quad \text{for all } r \in R(A)\,.$$

In (4.2), (4.4), and (4.5), S_σ denotes the map

$$S_\sigma(a_1 \otimes \cdots \otimes a_n) = (a_{\sigma^{-1}(1)} \otimes \cdots \otimes a_{\sigma^{-1}(n)})\,.$$

4.3 Examples from Chapter 2

Clearly,

$$(A_q^{2|0})^! \cong A_q^{0|2}\,.$$

[1] Here Π is the change of parity functor; in particular, starting from a purely even space A_1 we obtain, applying shriek, a purely odd space $\Pi(A_1^*)$.

One can look at the operations as natural lifts of standard functors from $\mathbb{K}$–mod to QA. Some lifts double, interchanged by the duality map !, as follows:

$$
\begin{array}{llllll}
\text{QA}: & \sim & \text{op} & ! & \circ \xleftrightarrow{\;!\;} \bullet & \otimes \xleftrightarrow{\;!\;} \underline{\otimes} \\
 & \downarrow & \downarrow & \downarrow & \searrow \;\; \swarrow & \searrow \;\; \swarrow \\
\mathbb{K}\text{–mod}: & \text{id} & \text{id} & * & \otimes & \oplus .
\end{array}
$$

For the sake of completeness, we list below the main interrelations between these functors.

4.4 Properties of ~

It is a covariant functor QA $\to$ QA; the canonical map $\tilde{A} \to A$ is a functor morphism $\tilde{} \to \mathrm{id}$. Moreover,

$$(\tilde{A})^{\mathrm{op}} = \widetilde{A^{\mathrm{op}}} = \tilde{A}\,;$$

$$\widetilde{(A^!)} = \tilde{A}^* := \bigoplus_i (\tilde{A}_i)^*\,, \qquad \text{more precisely,} \bigoplus_i \left(\Pi(\tilde{A}_i)\right)^*\,;$$

$$(\tilde{A})^! = \mathbb{K} \oplus A_1^*\,, \qquad \text{more precisely,} \, \mathbb{K} \oplus \Pi(A_1^*)\,.$$

Finally,

$$\widetilde{A \circ B} = \widetilde{A \bullet B} = \tilde{A} \circ \tilde{B} = \tilde{A} \bullet \tilde{B} = \text{tensor algebra of } A_1 \otimes B_1$$

$$\widetilde{A \otimes B} = \widetilde{A \,\underline{\otimes}\, B} = \text{tensor algebra of } A_1 \oplus B_1\,.$$

4.5 Properties of op

The map $A \to A^{\mathrm{op}}$, $f \mapsto f^{\mathrm{op}}$, where $(f^{\mathrm{op}})_1 = f_1$, is a covariant involution of QA. Denote by A° the ring, coinciding with A as a linear space, with “reversed multiplication”:

$$f * g(\text{in } A^\circ) = gf(\text{in } A)\,.$$

Then the map

$$\tau : \tilde{A} \to \tilde{A}\,, \quad \tau(a_1 \cdots a_n) = a_n \cdots a_1$$

induces an isomorphism $A^{\mathrm{op}} \xrightarrow{\sim} A^{\circ}$. Furthermore, there are functional identifications

$$(A^{\mathrm{op}})^! = \left(A^!\right)^{\mathrm{op}}, \quad (A * B)^{\mathrm{op}} = A^{\mathrm{op}} * B^{\mathrm{op}} ,$$

where $*$ is one of the products (4.4)–(4.7).

4.6 Properties of !

The dualization functor $A \to A^!$, $f \mapsto f^!$, where $f_1^! = f_1^*\colon B_1^* \to A_1^*$, is a contravariant quasi-involution of QA: indeed, $^{!!}$ is equivalent to id. There are natural identifications

$$(A \circ B)^! = A^! \bullet B^! ; \qquad (A \bullet B)^! = A^! \circ B^! ;$$
$$(A \otimes B)^! = A^! \underline{\otimes} B^! ; \qquad (A \underline{\otimes} B)^! = A^! \otimes B^! .$$

4.7 Properties of Products

The multiplications (4.4)–(4.7) with the associativity and commutativity morphisms, which are evident on components of degree 1, define on QA four different structures of tensor categories (for a more detailed discussion, see [16] and Chapter 13). All these tensor categories have unit objects; these are

$$\begin{aligned} &K = \mathbb{K}[\varepsilon] , \text{ where } \varepsilon^2 = 0 , && \text{for } \circ ; \\ &L = \mathbb{K}[t] = K^! && \text{for } \bullet ; \\ &\mathbb{K} && \text{for } \otimes \text{ and } \underline{\otimes} . \end{aligned}$$

However, in the category of "quantum linear spaces" $\mathrm{QA}^{\mathrm{op}}$, the products $\bullet$ and $\circ$ are analogues of the tensor product while $\otimes$ and $\underline{\otimes}$ correspond to the direct sum.

We also have functorial homomorphisms:

$$A \bullet B \xrightarrow{\alpha} A \circ B \xrightarrow{\beta} A \otimes B ,$$

where

$$\alpha_1 = \mathrm{id}\colon A_1 \otimes B_1 \to A_1 \otimes B_1 ;$$
$$\beta(a \otimes b) = a \otimes b \quad \text{for any } a \in A_1, b \in B_1 .$$

Note that β is *not* a morphism in QA since it doubles the degree (if, as usual, we agree that $(A \otimes B)_n = \bigoplus A_i \otimes B_{n-i}$).

4.8 Lemma

The mapping β induces an isomorphism of rings

$$A \circ B \xrightarrow{\sim} \sum_{n=0}^{\infty} A_n \otimes B_n \subset A \otimes B .$$

Proof. We just check that, in the notation of Section 4.2,

$$R_n(A \circ B) = S_\sigma\big(R_{2n}(A \otimes B) \cap (A_1^{\otimes n} \otimes B_1^{\otimes n})\big) ,$$

where σ is the permutation transforming $A_1^{\otimes n} \otimes B_1^{\otimes n}$ into $(A_1 \otimes B_1)^{\otimes n}$, and conserving the relative order of A- and B-factors. But from (4.8), we see that

$$R_n(A \circ B) = \sum_{i=1}^{n} (A_1 \otimes B_1)^{\otimes n} \otimes S_{(23)}\big(R(A) \otimes B_1^{\otimes 2} + A_1^{\otimes 2} \otimes R(B)\big) \otimes (A_1 \otimes B_1)^{\otimes (n-2-i)} ;$$

$$R_{2n}(A \otimes B) \cap A_1^{\otimes n} \otimes B_1^{\otimes n}$$
$$= \sum_{i=1}^{n} \big((A_1 \oplus B_1)^{\otimes 2i} \big(R(A) \oplus [A_1, B_1] \oplus R(B)\big) \otimes (A_1 \oplus B_1)^{\otimes 2n-2i-2}\big) \cap (A_1^{\otimes n} \otimes B_1^{\otimes n})$$

and a reshuffling of factors proves the lemma. □

4.9 Quantum Symmetric Power

For any $A \in \mathrm{Ob(QA)}$, define

$$A^{(d)} := \bigoplus_{i=0}^{\infty} A_{id} . \tag{4.9}$$

By analogy with the commutative polynomial case, we call $A^{(d)}$ (the ring of function on) the dth symmetric power of (the quantum space defined by) A.

4.10 Proposition

The algebra $A^{(d)}$ is quadratic.

Proof. Follows easily from (4.8). □

Backelin and Fröberg [5] proved, in fact, that if A is a graded algebra generated by a finite-dimensional space A_1, and if the ideal of relations is generated by its components of degree $\leq r$, then the same is true for $A^{(d)}$ with $\lfloor 2 + (r-2)/d \rfloor$ instead of r. Hence, any finitely presented graded algebra gives rise to a quadratic algebra under an operation $A \to A^{(d)}$. In the commutative case, this operation does not change $\operatorname{Proj} A$. Therefore *quadratic algebras essentially exhaust the range of projective algebraic geometry.*

Remark. In fact, by a result of Backelin [4], for high enough d and A commutative, $A^{(d)}$ is even Koszul.

4.11 Quantum Exterior Power

Since $S^\bullet(V)^! = \bigwedge^\bullet(\Pi(V^*))$, we can call $A^{!(d)}$ the dth exterior power of the dual space of A. However, it is better to reinterpret ! using the lessons of superalgebras: it is a functor which combines *dualization and parity change.* Without imposing some additional structures, we cannot define these two constructions separately.

4.12 Summary

Considering $\mathrm{QA}^{\mathrm{op}}$ as the category of "quantum linear spaces," we use the following analogues:

$\circ$: tensor product of quantum spaces;
$\otimes$: direct sum of quantum spaces;
!: dualization + parity change;
(d): dth symmetric power;
$(!d)$ or (Λd): dth exterior power.

Observe that the first four of the above operations are insufficient to describe both symmetric and exterior algebras of the *same* space. For more examples of similar type of subtleties, see [32].

Chapter 5
Quantum Matrix Spaces. I. Categorical Viewpoint

5.1 Motivation

Let U, V, T be finite-dimensional linear spaces. Then we have natural maps:

$$\mathrm{Hom}(U,V)\otimes U \to V : f\otimes u \mapsto f(u) \; ; \tag{5.1}$$
$$\mathrm{Hom}(V,T)\otimes \mathrm{Hom}(U,V) \to \mathrm{Hom}(U,T) : f\otimes g \mapsto fg \; , \tag{5.2}$$

with well-known universality properties.

Passing to the rings of polynomial functions $A(V) = S(V^*)$ we get dual maps

$$A(V) \to A\big(\mathrm{Hom}(U,V)\big)\circ A(U) \; ; \tag{5.3}$$
$$A\big(\mathrm{Hom}(U,T)\big) \to A\big(\mathrm{Hom}(V,T)\big)\circ A\big(\mathrm{Hom}(U,V)\big) \; . \tag{5.4}$$

We show that in QA, we have natural analogues of the maps (5.3) and (5.4), if we *define*

$$A\big(\mathrm{Hom}(U,V)\big) \; := \; A(U)^! \bullet A(V) \; . \tag{5.5}$$

Quadratic algebras $A^! \bullet B$ thus play the role of *quantum matrix spaces* (see, however, a warning at the end of this section).

Here we clarify the categorical content of this construction, while in the next chapters, we investigate quantum semigroups $A^! \bullet A$ and the problem of their extension to quantum groups using coordinates.

5.2 Theorem

There is a functorial isomorphism

Y. I. Manin, *Quantum Groups and Noncommutative Geometry*,
CRM Short Courses, https://doi.org/10.1007/978-3-319-97987-8_5

$$\mathrm{Hom}(A \bullet B, C) = \mathrm{Hom}(A, B^! \circ C)$$

identifying a map $f : A_1 \otimes B_1 \to C_1$ *with a map* $g : A_1 \to B_1^* \otimes C_1$, *if*

$$\langle g(a) \mid b \rangle = f(a \otimes b) \quad \text{for } a \in A_1, b \in B_1 \tag{5.6}$$

where the left-hand side denotes contraction with respect to B_1.

Corollary. *The pair* (QA, •) *is a (nonadditive) tensor category with internal* <u>Hom</u> *and unit object* $K = \mathbb{K}[\varepsilon]$, *where* $\varepsilon^2 = 0$.

Proof. We must check that if f, g are related as in (5.6), the following conditions are equivalent:

$$(f \otimes f)S_{(23)}\big(R(A) \otimes R(B)\big) \subset R(C) \, ,$$
$$(g \otimes g)R(A) \subset S_{(23)}\big(R(B)^{\perp} \otimes C_1^{\otimes 2} + B_1^{*\otimes 2} \otimes R(C)\big) \, .$$

Indeed, they are respectively equivalent to:

$$\big\langle R(C)^{\perp} \mid (f \otimes f)S_{(23)}\big(R(A) \otimes R(B)\big)\big\rangle = 0 \qquad \text{(contraction w.r.t. } C_1 \otimes C_1) \, ,$$

$$\langle R(B) \otimes R(C)^{\perp} \mid (g \otimes g)R(A) \rangle = 0 \qquad \text{(contraction w.r.t. } C_1 \otimes C_1 \otimes C_1^* \otimes C_1^*) \, .$$

Each of these last orthogonality relations means that if we start with an element of $R(A)$, apply $g \otimes g$ to it and then contract consecutively with arbitrary elements of $R(B)$ and $R(C)^{\perp}$, we get zero. □

5.3 Internal <u>Hom</u> Functor

Following the general formalism of tensor categories [16], we define

$$\underline{\mathrm{Hom}}(B, C) := B^! \circ C \, .$$

In particular,

$$B^! = \underline{\mathrm{Hom}}(B, K^!) \, .$$

Therefore, it is *not* the standard dualization in (QA, •) which would be

$$\check{B} = \mathrm{Hom}(B, K) = \mathbb{K} \otimes B_1^* \, ,$$

hardly a very interesting object.

By the general properties of Hom, the following natural maps are defined:

$$\beta\colon \underline{\mathrm{Hom}}(B,C)\bullet B\to C \tag{5.7}$$

$$\mu\colon \underline{\mathrm{Hom}}(C,D)\bullet \underline{\mathrm{Hom}}(B,C)\to \underline{\mathrm{Hom}}(B,D)\,, \tag{5.8}$$

with obvious associativity properties.

The map (5.7) has the following *universality property*: for any morphism

$$f\colon A\bullet B\to C \quad \text{in QA}\,,$$

there exists a unique morphism $g\colon A\to \underline{\mathrm{Hom}}(B,C)$ making the following diagram commutative:

$$\begin{array}{ccc} \underline{\mathrm{Hom}}(B,C)\bullet B & \xrightarrow{\beta} & C\,. \\ {\scriptstyle g\bullet \mathrm{id}}\uparrow & \nearrow{\scriptstyle f} & \\ A\bullet B & & \end{array}$$

This map is, actually, just the identification defined in Theorem 5.2. One can then construct (5.8) iterating (5.7).

5.4 Internal hom Functor

Comparing (5.7) and (5.8) with (5.3) and (5.4), we see, sadly, that $\underline{\mathrm{Hom}}(B,C)$ is *not* the quantum matrix space we are looking for. In fact it is *dual* to it with respect to !.

Define

$$\underline{\mathrm{hom}}(B,C) := \underline{\mathrm{Hom}}(B^!,C^!)^! = B^!\bullet C\,.$$

Write (5.7) and (5.8) for $B^!, C^!, D^!$ and apply !. We get morphisms

$$\delta\colon C\to \underline{\mathrm{hom}}(B,C)\circ B\,, \tag{5.9}$$

$$\Delta\colon \underline{\mathrm{hom}}(B,D)\to \underline{\mathrm{hom}}(C,D)\circ \underline{\mathrm{hom}}(B,C)\,, \tag{5.10}$$

defined by a universality property dual to the one in Section 5.3.

5.5 Theorem

(a) *For any morphism* $g: C \to A \circ B$ in QA *there exists a unique morphism* $f: \underline{\hom}(B, C) \to A$ *such that* $g = (f \circ \mathrm{id}_b)\delta$.
(b) *Let* $B = C = D$ in (5.9) *and* (5.10). *Then*

$$E = \underline{\mathrm{end}}(B) = \underline{\hom}(B, B) = B^! \bullet B$$

becomes a bialgebra with respect to the following data:

$$\begin{aligned} m_E &= \text{multiplication in } E\ ; \\ \Delta_E &= \text{composition of the maps } E \xrightarrow{(5.10)} E \circ E \to E \otimes E\ ; \\ \eta&: \mathbb{K} = E_0 \xrightarrow{\overset{\mathrm{id}}{\sim}} \mathbb{K}\ ; \\ \varepsilon&: E_1 = B_1^* \otimes B_1 \to \mathbb{K} \quad \text{the standard pairing}\ . \end{aligned}$$

(c) *The map* $\delta: B \to E \circ B$, *see* (5.9), *induces on* B (and on *all homogeneous components* B_i of B) *a structure of left* E*-comodule.*

Part (a) of this theorem is again a restatement of Theorem 5.2.

We prove parts (b) and (c) in the next chapter using coordinate language, and return to the categorical picture in Chapter 13.

5.6 Warning

If we apply our constructions to the usual linear spaces, i.e., polynomial rings, we do *not* obtain matrices with commuting entries!

Indeed, $\underline{\mathrm{end}}\,\mathbb{K}[x_1, \ldots, x_n]$ is a very noncommutative ring: it is defined only by half of the necessary commutation relations. In order to get the other half, we can impose the same commutation relations upon the transposed matrix, as in the proof of Theorem 2.4. For details, see the next chapter.

Chapter 6
Quantum Matrix Spaces. II. Coordinate Approach

6.1 A Quantum Space

Consider a general quadratic algebra

$$A = \mathbb{K}\langle \tilde{x}_1, \ldots, \tilde{x}_n \rangle / (r_\alpha),$$

where $\mathbb{K}\langle \tilde{x} \rangle$ means a free associative algebra generated by the $\tilde{x}_j$, and

$$r_\alpha = r_\alpha(\tilde{x}) = \sum_{i,j} c_\alpha^{ij} \tilde{x}_i \tilde{x}_j \quad \text{for } \alpha = 1, \ldots, m \, , \tag{6.1}$$

are linearly independent elements of $\mathbb{K}\langle \tilde{x}_1, \ldots, \tilde{x}_n \rangle_2$. We define $R := (r_\alpha)$ and $x_i := \tilde{x}_i \bmod R$; we also denote by R the set of relations in any algebra to appear later.

6.2 The Dual Space

This is defined by the algebra

$$A^! = \mathbb{K}\langle \tilde{x}^1, \ldots, \tilde{x}^n \rangle / (r^\beta),$$

where

$$r^\beta = \sum_{k,l} c_{kl}^\beta \tilde{x}^k \tilde{x}^l \quad \text{for } \beta = 1, \ldots, n^2 - m \, , \tag{6.2}$$

and where $\langle \tilde{x}^i \mid \tilde{x}_j \rangle = \delta^i_j$ and $\bigoplus \mathbb{K} r^\beta = (\bigoplus \mathbb{K} r_\alpha)^\perp$, i.e.,

Y. I. Manin, *Quantum Groups and Noncommutative Geometry*,
CRM Short Courses, https://doi.org/10.1007/978-3-319-97987-8_6

$$\langle r^\beta \mid r_\alpha \rangle = \sum_{i,j} c_{ij}^\beta c_\alpha^{ij} = 0 \,. \tag{6.3}$$

We set $x^i := \tilde{x}^i \bmod R$.

6.3 The Matrix Space $\underline{\mathrm{end}}(A)$

We have

$$\begin{aligned} \underline{\mathrm{end}}(A) &= A^! \bullet A = \mathbb{K}\langle \tilde{z}_i^k \mid 1 \le i,k \le n\rangle/(r_\alpha^\beta)\,; \\ \tilde{z}_i^k &= \tilde{x}^k \otimes \tilde{x}_i\,; \quad z_i^k = \tilde{z}_i^k \bmod R\,; \\ r_\alpha^\beta &= S_{(23)}\big(r^\beta \otimes r_\alpha\big) = \sum_{i,j,k,l} c_\alpha^{ij} c_{kl}^\beta \tilde{z}_i^k \tilde{z}_j^l\,. \end{aligned} \tag{6.4}$$

6.4 Coaction

This is an algebra map (a morphism in QA)

$$\delta_A \colon A \to \underline{\mathrm{end}}(A) \circ A \subset \underline{\mathrm{end}}(A) \otimes A\,, \quad \delta_A(x_i) = \sum_k z_i^k \otimes x_k\,. \tag{6.5}$$

6.5 Lemma

The coaction (6.5) *is well-defined. Moreover, let*

$$\tilde{\delta} \colon \mathbb{K}\langle \tilde{x}_1, \ldots, \tilde{x}_n\rangle \to \mathbb{K}\langle \tilde{z}_i^k\rangle \otimes \mathbb{K}\langle \tilde{x}_j\rangle$$

be defined by the same formula (6.5). *Then for some* $s_\beta(\tilde{x}\rangle \in \mathbb{K}\langle \tilde{x}\rangle$ *and* $s_\alpha^\beta \in \mathbb{K}\langle \tilde{z}\rangle$, *we have*

$$\tilde{\delta}(r_\alpha) = \sum_\beta r_\alpha^\beta(\tilde{z}) \otimes s_\beta(\tilde{x}) + \sum_{\alpha'} s_\alpha^{\alpha'}(\tilde{z}) \otimes r_{\alpha'}(\tilde{x})\,. \tag{6.6}$$

Proof. Choose $n^2 - m$ quadratic forms $s_\beta = s_\beta(\tilde{x}) \in \mathbb{K}\langle \tilde{x}\rangle$ in such a way that $\langle r^\beta \mid s_\gamma\rangle = \delta_\gamma^\beta$. Then the set $\{r_\alpha, s_\beta\}$ forms a basis of $\mathbb{K}\langle \tilde{x}\rangle_2$. We affirm that

$$\tilde{x}_k \tilde{x}_l = \sum_\beta c_{kl}^\beta s_\beta + \sum_\alpha e_{kl}^\alpha r_\alpha$$

for some c_{kl}^{β} from (6.2) and some constants e_{kl}^{α}. This is checked by taking scalar products with r^{β}. Now,

$$\begin{aligned}\tilde{\delta}(r_\alpha) &= \sum_{i,j} c_\alpha^{ij} \Big(\sum_k \tilde{z}_i^k \otimes \tilde{x}_k\Big)\Big(\sum_l \tilde{z}_j^l \otimes \tilde{x}_l\Big) \\ &= \sum_{i,j,k,l} c_\alpha^{ij} \tilde{z}_i^k \tilde{z}_j^l \otimes \Big(\sum_\beta c_{kl}^\beta s_\beta + \sum_{\alpha'} e_{kl}^{\alpha'} r_{\alpha'}\Big) \\ &= \sum_{\beta,i,j,k,l} c_\alpha^{ij} c_{kl}^\beta \tilde{z}_i^k \tilde{z}_j^l \otimes s_\beta(\tilde{x}) + \sum_{\alpha',i,j,k,l} c_\alpha^{ij} e_{kl}^{\alpha'} \tilde{z}_i^k \tilde{z}_j^l \otimes r_{\alpha'}(\tilde{x})\end{aligned}$$

which proves (6.6) in view of (6.4). □

6.6 Lemma on Universality

For any $\mathbb{K}$- algebra B and morphism $\delta\colon A \to B \otimes A$ with $\delta(A_1) \subset B \otimes A_1$, there exists a unique morphism $\gamma\colon \underline{\text{end}}(A) \to B$ such that the diagram

$$\begin{array}{ccc} A & \xrightarrow{\delta} & B \otimes A \\ & \searrow_{\delta_A} & \uparrow{\scriptstyle \gamma\otimes \mathrm{id}} \\ & & \underline{\text{end}}(A) \otimes A \end{array}$$

commutes. If, moreover,

$$\delta\colon A \to B \circ A \to B \otimes A$$

is a morphism of quadratic algebras, then γ is also a morphism in QA.

Proof. Let $v_i^k \in B$ be defined from the formula $\delta(x_i) = \sum v_i^k \otimes x_k$. Plugging x_i and v_i^k into the universal formula (6.6) instead of $\tilde{x}_i$ and $\tilde{z}_i^k$, we get

$$0 = \delta\big(r_\alpha(x_i)\big) = \sum_\beta r_\alpha^\beta(v) \otimes s_\beta(x)\,.$$

Hence, $r_\alpha^\beta(v) = 0$ since the elements $s_\beta(x)$ are linearly independent in A_2. Therefore, we can define $\gamma\colon \underline{\text{end}}(A) \to B$ by setting $\gamma(z_i^k) = v_i^k$. The rest is clear. □

6.7 The Diagonal Map

Using universality, we can now *define* the diagonal map $\Delta_A : E \to E \circ E \to E \otimes E$, where $E = \underline{\mathrm{end}}(A)$, by the commutative diagram:

$$A \xrightarrow{\ \delta\ } E \otimes A \xrightarrow{\ \mathrm{id}\otimes\delta_A\ } E \otimes E \otimes A\,.$$
$$A \xrightarrow{\ \delta_A\ } E \otimes A \xrightarrow{\ \Delta_A\otimes\mathrm{id}\ } E \otimes E \otimes A$$

In this way, the coassociativity for δ_a, see Section 3.8, becomes evident.

Applying morphisms in this diagram to x_i, we get

$$\Delta\left(z_i^k\right) = \sum_j z_i^j \otimes z_j^k\ ,$$

or, as we wrote in Section 3.6, $\Delta(Z) = Z \otimes Z$ for $Z = \left(z_i^k\right)$.

6.8 The End of the Proof of Theorem 5.5

We have already defined most of the data, describing E and the action of E on A, and checked that most of the axioms are satisfied. We have not mentioned the counit. Define

$$\varepsilon\left(z_i^k\right) := \delta_i^k\ .$$

We have $r_\alpha^\beta\left(\delta_i^k\right) = \delta_i^k$ in view of condition (6.3) so that the relations in E are not violated. The counit axioms for Δ_A and δ_A are now evidently satisfied. □

For future constructions it is useful to keep in mind an analogue of Lemma 6.5 for Δ_A: *if* $\tilde{\Delta}(\widetilde{Z}) = \widetilde{Z} \otimes \widetilde{Z}$, *then*

$$\tilde{\Delta}(r_\alpha^\beta) = \sum_{\gamma,\delta} r_\gamma^\delta(\tilde{z}) \otimes s_{\alpha\delta}^{\beta\gamma}(\tilde{z}) + \sum_{\gamma,\delta} t_{\alpha\delta}^{\beta\gamma}(\tilde{z}) \otimes r_\gamma^\delta(\tilde{z}) \tag{6.7}$$

for some $s, t \in \mathbb{K}\langle \tilde{z}_i^k \mid 1 \le i, k \le n\rangle$.

(This analogue of Lemma 6.5 follows from what we have already proved, but can also be proved directly.)

We use formulas of this kind in order to check that an ideal in a bialgebra is, actually, a coideal, and therefore it can be factored out consistently with the codiagonal map.

Let us now discuss the functorial behavior of $\underline{\mathrm{end}}(A)$.

6.9 Dualization

In commutative geometry, $\mathrm{end}\, L \cong \mathrm{end}\, L^*$ canonically, as linear spaces, but multiplication is reversed. Exactly the same holds in our situation.

6.10 Theorem

There is a canonical isomorphism of bialgebras

$$\tau_A \colon (\underline{\mathrm{end}}(A), m_A, \Delta_A) \xrightarrow{\sim} (\underline{\mathrm{end}}(A^!), m_{A^!}, \Delta_{A^!}^{\mathrm{op}})$$

coinciding with $S_{(12)} \colon A_1^* \otimes A_1 \to A_1 \otimes A_1^*$ *on 1-components.*

Proof. We work it out in coordinates. Put $\check{x}_i := \tilde{x}^i$ and $\check{x}^i := \tilde{x}_i$ thus identifying $A^{!!} = A$. Then

$$A^! = \mathbb{K}\langle \check{x} \rangle / (\check{r}_\beta)$$

$$\check{r}_\beta = \sum_{i,j} \check{c}_\beta^{ij}\, \check{x}_i \check{x}_j\ , \quad \check{c}_\beta^{ij} = c_{ij}^\beta \qquad \text{for } \beta = 1, \ldots, n^2 - m\ .$$

Similarly,

$$\check{r}^\alpha = \sum_{kl} \check{c}_{kl}^\alpha \check{x}^k \check{x}^l\ , \quad \check{c}_{kl}^\alpha = c_\alpha^{kl} \qquad \text{for } \alpha = 1, \ldots, m\ .$$

Therefore,

$$\underline{\mathrm{end}}(A^!) = A \bullet A^! = \mathbb{K}\langle \check{z}_i^k \mid 1 \le i, k \le n \rangle / (\check{r}_\beta^\alpha)\ ,$$

where

$$\check{r}_\beta^\alpha = S_{(23)}(\check{r}^\alpha \otimes \check{r}_\beta) = \sum \check{c}_\beta^{ij} \check{c}_{kl}^\alpha \check{z}_i^k \check{z}_j^l\ . \tag{6.8}$$

Let $\widetilde{Z} := \langle \tilde{z}_i^k \rangle$ and $\check{Z} := \langle \check{z}_i^k \rangle$. Comparing relations (6.8) with (6.4), we see that the map $\tilde{\tau} \colon \widetilde{Z} \to \check{Z}^{\mathrm{t}}$ satisfies $\tilde{\tau}(r_\alpha^\beta) = \check{r}_\beta^\alpha$, and therefore defines an algebra isomorphism $\underline{\mathrm{end}}(A) \xrightarrow{\sim} \underline{\mathrm{end}}(A^!)$. In other words, the relations $\check{r}$ can be obtained from the relations r by transposing Z.

Finally,

$$\Delta^{\mathrm{op}}_{A^!}(\check{Z}^{\mathrm{t}}) = \check{Z}^{\mathrm{t}} \otimes \check{Z}^{\mathrm{t}}$$

in view of Proposition 3.7(a), since $\Delta_{A^!}(\check{Z}) = \check{Z} \otimes \check{Z}$. Hence $\tilde{\tau}$ induces the bialgebra isomorphism.

Clearly, $\tilde{\tau}$ coincides with $S_{(12)}$ on generators. Let us write formulas of the coordinate change for future use. If

$$x' = \begin{pmatrix} x'_1 \\ \vdots \\ x'_n \end{pmatrix} = U \begin{pmatrix} x_1 \\ \vdots \\ x_n \end{pmatrix}$$

in A_1, where $U \in \mathrm{GL}(n, \mathbb{K})$, then

$$\check{x}' = (U^{\mathrm{t}})^{-1} x \ , \quad Z' = UZU^{-1} \ , \quad \check{Z}' = (U^{\mathrm{t}})^{-1} \check{Z} U^{\mathrm{t}} \ , \tag{6.9}$$

and we see that τ_A is well-defined. □

6.11 The Opposite Bialgebra

In coordinates, we have

$$A^{\mathrm{op}} := \mathbb{K}\langle \mathring{x}_1, \ldots, \mathring{x}_n \rangle / (\mathring{r}_\alpha) \ , \quad \text{where } \mathring{r}_\alpha = \sum \mathring{c}^{ij}_\alpha \mathring{x}_i \mathring{x}_j \text{ and } \mathring{c}^{ij}_\alpha = c^{ji}_\alpha \ .$$

Therefore, in self-evident notation,

$$\underline{\mathrm{end}}(A^{\mathrm{op}}) = \mathbb{K}\langle \mathring{z}^k_i \mid 1 \le i, k \le n \rangle / (\mathring{r}_\alpha) \ .$$

6.12 Theorem

The map $\tilde{\rho}\colon \widetilde{Z} \to \mathring{Z}$, where $\tilde{\rho}(fg) = \tilde{\rho}(g)\tilde{\rho}(f)$, satisfies $\tilde{\rho}(r^\beta_\alpha) = \mathring{r}^\beta_\alpha$, and therefore $\tilde{\rho}$ induces a canonical bialgebra isomorphism

$$\rho_A\colon (\underline{\mathrm{end}}(A), m_A, \Delta_A) \to (\underline{\mathrm{end}}(A^{\mathrm{op}}), m^{\mathrm{op}}_{A^{\mathrm{op}}}, \Delta_{A^{\mathrm{op}}}) \ .$$

Proof. Clear. □

6.13 The Symmetric Power

As explained earlier, we may consider A_1 as the tautological (co)module over $\underline{\mathrm{end}}(A)$ and A_d as the dth symmetric power of A_1. On the other hand, it is more suggestive to imagine the *whole quantum space* A as the tautological $\underline{\mathrm{end}}(A)$-module and $A^{(d)}$ as its dth symmetric power. Then we have the coaction

$$\delta_A|_{A^{(d)}}: A^{(d)} \to \underline{\mathrm{end}}(A) \otimes A^{(d)} .$$

Since, clearly, $\delta_A(A_1^{(d)}) \subset \underline{\mathrm{end}}(A) \otimes A_1^{(d)}$, the map $\delta_A|_{A^{(d)}}$ is induced in view of Lemma 6.6 by a morphism (the through map)

$$\sigma^{(d)}: \underline{\mathrm{end}}\big(A^{(d)}\big) \to \underline{\mathrm{end}}(A)^{(d)} \to \underline{\mathrm{end}}(A)$$

which is the third version of the *quantum symmetric power* (see the discussion in Section 3.10).

Chapter 7
Adding Missing Relations

7.1 Example

We start with the commutative polynomial algebra

$$A = \mathbb{K}\langle \tilde{x}_1, \dots, \tilde{x}_n \rangle / (\tilde{x}_i \tilde{x}_j - \tilde{x}_j \tilde{x}_i) .$$

Then

$$A^! = \mathbb{K}\langle \tilde{x}^1, \dots, \tilde{x}^n \rangle / (\tilde{x}^k \tilde{x}^l + \tilde{x}^l \tilde{x}^k) ,$$

and

$$\begin{aligned}
\underline{\mathrm{end}}(A) &= \mathbb{K}\langle \tilde{z}_i^k \mid 1 \le i, k \le n \rangle \Big/ \left(r_{(ij)}^{(kl)} \right) , \\
r_{(ij)}^{(kl)} &= S_{(23)}\big((\tilde{x}^k \tilde{x}^l + \tilde{x}^l \tilde{x}^k)(\tilde{x}_i \tilde{x}_j - \tilde{x}_j \tilde{x}_i) \big) \\
&= \tilde{z}_i^k \tilde{z}_j^l - \tilde{z}_j^k \tilde{z}_i^l + \tilde{z}_i^l \tilde{z}_j^k - \tilde{z}_j^l \tilde{z}_i^k \\
&= [\tilde{z}_i^k , \tilde{z}_j^l] + [\tilde{z}_i^l , \tilde{z}_j^k] .
\end{aligned} \tag{7.1}$$

In all, we have $\frac{1}{2}n(n-1) \cdot \frac{1}{2}n(n+1)$ relations for n^2 matrix entries z_i^k, and hence, for $n > 1$, the ring $\underline{\mathrm{end}}(A)$ is highly noncommutative. But if we add $\frac{1}{4}n^2(n^2-1)$ more relations, requiring (7.1) for the transposed matrix Z^{t}, we get $\mathbb{K}[z_i^j]_{i,j=1}^n$, the common ring of functions on the space of matrices. In fact, (7.1) means that for an arbitrary 2×2 submatrix of Z, the commutators of its diagonal elements mod R have opposite signs:

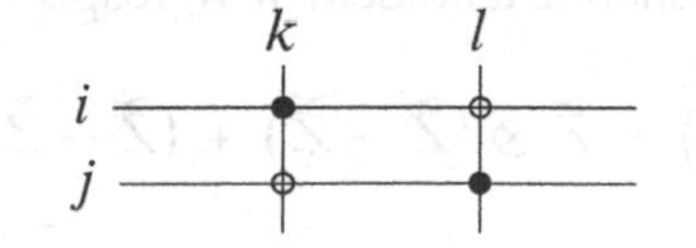

Y. I. Manin, *Quantum Groups and Noncommutative Geometry*,
CRM Short Courses, https://doi.org/10.1007/978-3-319-97987-8_7

But in the transposed matrix, just one of the commutators changes sign. Therefore, all of them must vanish.

7.2 General Case

In the proof of Theorem 6.10 we saw that the relations r for the transposed matrix essentially coincide with the relations $\check{r}$ defining $\operatorname{end}(A^!)$. Therefore we may combine the generators $z_i^k, \check{z}_i^k$ and relations $r, \check{r}, Z - \check{Z}$ to achieve the necessary effect. The result may, however, depend on the choice of coordinates x_i in A. Looking at the coordinate change in the formulas (6.9) we see that the relations $Z - \check{Z}$ are invariant (up to a linear transformation) only with respect to the *orthogonal* coordinate changes $x' = Ux$, where $UU^{\mathrm{t}} = E$. This means that we implicitly fix an orthogonal form $g = x_1^2 + \cdots + x_n^2 \in S^2(A_1)$, i.e., identify A_1 and $A_1^!$ "in a symmetric way." (Hence, e.g., $\mathrm{GL}_q(2)$ from Chapter 2 is a "cryptorthogonal" group! Of course, instead of having a fixed g, we may equip A_1 with a marked basis.) Formally, we write

$$
\begin{aligned}
\underline{\mathrm{e}}(A, g) &= \mathbb{K}\langle \tilde{z}_i^k, \check{z}_i^k \mid 1 \le i, k \le n \rangle / (r(\widetilde{Z}), \check{r}(\check{Z}), \widetilde{Z} - \check{Z}) \\
&\cong \mathbb{K}\langle \tilde{z}_i^k \mid 1 \le i, k \le n \rangle / \big(r(\widetilde{Z}), \check{r}(\widetilde{Z})\big) \\
&\cong \mathbb{K}\langle \check{z}_i^k \mid 1 \le i, k \le n \rangle / \big(r(\check{Z}), \check{r}(\check{Z})\big) .
\end{aligned}
$$

7.3 Theorem

(a) *The diagonal map* $\tilde{\Delta}\colon \widetilde{Z} \mapsto \widetilde{Z} \otimes \widetilde{Z}, \check{Z} \mapsto \check{Z} \otimes \check{Z}$ *descends to* $\underline{\mathrm{e}}(A, g)$; *the maps*

$$\delta(x_i) = \sum \tilde{z}_i^k \otimes \tilde{x}_k , \quad \delta(x^i) = \sum \check{z}_i^k \otimes \tilde{x}^k$$

define on A *and* $A^!$ *structures of left* $\underline{\mathrm{e}}(A, g)$*-comodules.*

(b) *The transposition map* $r\colon \widetilde{Z} \mapsto \widetilde{Z}^{\mathrm{t}}$ *defines an isomorphism of bialgebras* $(\underline{\mathrm{e}}(A, g), m, \Delta) \xrightarrow{\sim} (\underline{\mathrm{e}}(A, g), m, \Delta^{\mathrm{op}})$.

Proof. Everything follows from the previous discussion. In particular, the entries of $r(\widetilde{Z}), \check{r}(\check{Z})$ and $\widetilde{Z} - \check{Z}$ generate a coideal with respect to $\tilde{\Delta}$ in view of (6.7) and

$$\tilde{\Delta}(\widetilde{Z} - \check{Z}) = \widetilde{Z} \otimes (\widetilde{Z} - \check{Z}) + (\widetilde{Z} - \check{Z}) \otimes \check{Z} .$$

We leave the rest to the reader. □

Of course, this construction is specially interesting if $R(A)$ bears some relation to g. Let us consider three examples.

7.4 Quantum Conformal Group

Let $A = \mathbb{K}\langle x\rangle/(\sum x_i^2)$, i.e., $R(A) = \mathbb{K}g$, where $n \geq 2$. In notation of Chapter 6, we have $m = 1$, $c_1^{ij} = \delta^{ij}$. Equation (6.3) shows that the matrices (c_{ij}^{β}) form a basis of traceless matrices.

Hence, the algebra $A^!$ can be defined by the relations

$$\begin{aligned} x^i x^j &= 0 && \text{for } i \neq j\ , \\ (x^i)^2 &= (x^j)^2 && \text{for all } i, j\ . \end{aligned}$$

In particular, $\dim A_1^! = n$, $\dim A_2^! = 1$, $\dim A_n^! = 0$ if $n \geq 3$ (and $A^! = \mathbb{K}[x^1]$ if $n = 1$).

Furthermore, $\underline{\mathrm{end}}(A)$ is defined by the relations

$$\begin{aligned} \sum_k z_k^i z_k^j &= 0 && \text{for } i \neq j\ , \\ \left(z_i^i\right)^2 &= \left(z_j^j\right)^2 && \text{for all } i, j\ , \end{aligned}$$

or, briefly,

$$Z^{\mathrm{t}} Z = \text{scalar}\ .$$

Finally,

$$\underline{\mathrm{e}}(A, g) = \mathbb{K}\langle z_i^k \mid 1 \leq i, k \leq n\rangle/(Z^{\mathrm{t}} Z = \text{scalar}, Z Z^{\mathrm{t}} = \text{scalar})\ .$$

7.5 Case $R(A) = R(A)^{\perp}$ with Respect to g

In this case, the map $\tilde{g}\colon A_1 \to A_1^* = (A^!)_1$ induces an isomorphism $A \xrightarrow{\sim} A^!$, and therefore an isomorphism $\underline{\mathrm{end}}(A) \xrightarrow{\sim} \underline{\mathrm{end}}(A^!)\colon \tilde{Z} \mapsto \check{Z}$ (mod R). This means that $\underline{\mathrm{e}}(A, g) = \underline{\mathrm{end}}(A)$.

Ezra Getzler showed me a remarkable example of such a self-dual quadratic algebra, or rather of a bundle of such algebras over a Riemannian space. It is a graded bundle associated with a filtration of the sheaf of differential operators acting on a spinor bundle. (Actually, in this way one gets self-dual quadratic *super*algebras, cf. Chapters 12 and 13.)

7.6 Case $R(A) \oplus R(A)^{\perp} = A_1^{\otimes 2}$ with $\dim R(A) = n(n-1)/2$

This case is close to the commutative one, since the dimension of the space of quadratic relations for A (resp. $A^!$, resp. $\underline{e}(A, g)$) is the same as that for $\mathbb{K}[x_1, \ldots, x_n]$ (resp. Grassmann algebra in n indeterminates, resp. $\mathbb{K}[z_i^k]_{i,k=1}^n$). The deformations of polynomial algebras, considered by Drinfeld [21], belong to this class.

7.7 "Pseudosymmetric" Quantum Spaces

Consider an operator

$$R\colon A_1 \otimes A_1 \to A_1 \otimes A_1 \colon \quad \tilde{x}_i \otimes \tilde{x}_j \mapsto \sum_{k,l} r_{ij}^{kl}\, \tilde{x}_k \otimes \tilde{x}_l \,.$$

Assume that the following conditions hold:

(i) The relations $r_{(ij)} := \sum_{p,q} (\delta_{ij}^{pq} - r_{ij}^{pq}) \tilde{x}_p \otimes \tilde{x}_q$ for $1 \leq i < j \leq n$ are linearly independent.
(ii) The relations $r^{(kl)} := \sum_{s,t} (\delta_{st}^{kl} - r_{st}^{kl}) \tilde{x}^s \otimes \tilde{x}^t$ for $1 \leq k \leq l \leq n$ are linearly independent.
(iii) $(R^2)_{ij}^{kl} := \sum_{p,q} r_{ij}^{pq} r_{pq}^{kl} = \delta_{ij}^{kl}$ for $i < j$ and $k \leq l$.

Denote by A the quantum space defined by $r_{(ij)}$ and $g = \sum x_i^2$. Let $\widetilde{Z} \odot \widetilde{Z}$ be the $(n^2 \times n^2)$-matrix (usually called the *tensor product of matrices*)

$$(\widetilde{Z} \odot \widetilde{Z})_{ij}^{kl} = \tilde{z}_i^k \otimes \tilde{z}_j^l$$

and $(R)_{ij}^{kl} = r_{ij}^{kl}$. Then we have the following proposition.

7.8 Proposition

(a) *The algebra $A^!$ is defined by relations $r^{(kl)}$, and the relations in $\underline{\text{end}}(A)$ are generated by matrix elements $(I - R)\big(\widetilde{Z} \odot \widetilde{Z}\big)(I + R)$.*

(b) *If, in addition,* $R^{\mathrm{t}} = R$ *and* $R^2 = 1$*, then* $\underline{\mathrm{e}}(A, g)$ *is defined by the relations*

$$R(\widetilde{Z} \odot \widetilde{Z}) - (\widetilde{Z} \odot \widetilde{Z})R\,. \tag{7.2}$$

Proof. (a) Since $\dim\big(\bigoplus \mathbb{K}r_{(ij)}\big) = \frac{1}{2}n(n-1)$ and $\dim\big(\bigoplus \mathbb{K}r^{(kl)}\big) = \frac{1}{2}n(n+1)$, to check that $A^! = \mathbb{K}\langle \tilde{x}^i \rangle / (r^{(kl)})$ it suffices to prove that $\langle r^{(kl)} \mid r_{(ij)} \rangle = 0$. In fact, using (iii) we get:

$$\begin{aligned}\langle r^{(kl)} \mid r_{(ij)} \rangle = \sum_{s,t} (\delta^{kl}_{st} + r^{kl}_{st})(\delta^{st}_{ij} - r^{st}_{ij}) &= (I - R)(I + R)^{kl}_{ij} \\ &= \delta^{kl}_{ij} - (R^2)^{kl}_{ij} = 0\,.\end{aligned}$$

Now, $\underline{\mathrm{end}}(A)$ is defined by the relations

$$\begin{aligned}r^{(kl)}_{(ij)} = S_{(23)}\big(r^{(kl)} \otimes r_{(ij)}\big) &= \sum_{p,q,s,t} (\delta^{kl}_{st} + r^{kl}_{st})(\delta^{pq}_{ij} - r^{pq}_{ij}) \tilde{z}^s_p \otimes \tilde{z}^t_q \\ &= (I - R)(\widetilde{Z} \odot \widetilde{Z})(I + R)\,.\end{aligned} \tag{7.3}$$

(b) Writing relations (7.3) for $\widetilde{Z}^{\mathrm{t}}$ instead of $\widetilde{Z}$ and transposing, we find

$$(I + R^{\mathrm{t}})(\widetilde{Z} \odot \widetilde{Z})(I - R^{\mathrm{t}})\,. \tag{7.4}$$

If $R = R^{\mathrm{t}}$, then the relations (7.3) and (7.4) together are equivalent to

$$\{R(\widetilde{Z} \odot \widetilde{Z}) - (\widetilde{Z} \odot \widetilde{Z})R, R(\widetilde{Z} \odot \widetilde{Z})R\}\,, \tag{7.5}$$

which boils down to relations (7.2) for $R^2 = I$. □

Remark. One sees that if $R = -R^{\mathrm{t}}$, then $\underline{\mathrm{e}}(A, g) = \underline{\mathrm{end}}(A)$.

7.9 Proposition

For an arbitrary operator $R\colon A_1^{\otimes 2} \to A_1^{\otimes 2}$*, the matrix relation* (7.2) *defines a coideal with respect to* $\Delta(\widetilde{Z}) = \widetilde{Z} \otimes \widetilde{Z}$*, and hence defines a new bialgebra.*

Proof. Clearly,

$$\Delta(\widetilde{Z} \odot \widetilde{Z}) = S_{(23)}\big((\widetilde{Z} \odot \widetilde{Z}) \otimes (\widetilde{Z} \odot \widetilde{Z})\big)\,.$$

Therefore, if R commutes with $Z \odot Z$, it also commutes with $\Delta(Z \odot Z)$. □

Remark. In the QIST method, the relations (7.2) are used to define a quantum group by means of a (weak) Yang–Baxter operator R, cf. Chapter 12.

Chapter 8
From Semigroups to Groups

8.1 Motivation

In this chapter, we show that for quantum semigroups like $E = \underline{\mathrm{end}}(A)$ or $\underline{\mathrm{e}}(A, g)$ there exists a universal map $\gamma\colon E \to H$ into a Hopf algebra H. In view of Proposition 3.7, γ makes all multiplicative matrices in E invertible. Hence a natural idea is to add, formally, the necessary inverse matrices. The following construction suffices to treat $\underline{\mathrm{end}}(A)$ and $\underline{\mathrm{e}}(A, g)$.

8.2 Construction

Let E be a bialgebra generated by entries of a multiplicative matrix Z, and let $E := \mathbb{K}\langle\tilde{Z}\rangle/R_0$, where R_0 is the ideal of relations between matrix entries of Z, i.e., $Z = \tilde{Z} \bmod R_0$.

Let $\widetilde{Z}_0 = \widetilde{Z}$ and introduce a series of matrices $\widetilde{Z}_1, \widetilde{Z}_2, \dots$ in such a way that all entries of $\widetilde{Z}_0, \widetilde{Z}_1, \dots$ generate a free associative algebra. Denote by H the factor algebra of $\widetilde{H} = \mathbb{K}\langle\widetilde{Z}_0, \widetilde{Z}_1, \widetilde{Z}_2, \dots\rangle$ quotiented by the ideal generated by the following relations.[1]

$$R_k = \begin{cases} \text{elements of } R_0 \text{ written for } \widetilde{Z}_k \text{ instead of } \widetilde{Z} \\ \qquad\qquad\qquad\qquad \text{for } k \equiv 0 \pmod 2 \\ \text{elements of } R_0^{\mathrm{op}} \text{ written for } \widetilde{Z}_k \text{ instead of } \widetilde{Z} \\ \qquad\qquad\qquad\qquad \text{for } k \equiv 1 \pmod 2 \end{cases} \tag{8.1}$$

$$\widetilde{Z}_k\widetilde{Z}_{k+1} - I\,, \quad \widetilde{Z}_{k+1}\widetilde{Z}_k - I \qquad \text{for } k \equiv 0 \pmod 2\,, \tag{8.2}$$

[1] Hereafter $\mathbb{K}\langle Z_0, Z_1, \dots\rangle$ and similar expressions denote the algebra freely generated by the matrix entries.

Y. I. Manin, *Quantum Groups and Noncommutative Geometry*,
CRM Short Courses, https://doi.org/10.1007/978-3-319-97987-8_8

$$\widetilde{Z}_k^{\mathrm{t}} \widetilde{Z}_{k+1}^{\mathrm{t}} - I \,, \quad \widetilde{Z}_{k+1}^{\mathrm{t}} \widetilde{Z}_k^{\mathrm{t}} - I \qquad \text{for } k \equiv 1 \pmod 2 \,. \tag{8.3}$$

Define $\gamma\colon E \to H$ by setting $\gamma(Z) = \widetilde{Z}_0 \pmod R$.

8.3 Theorem

(a) *Relations* (8.1)–(8.3) *generate a coideal* $\widetilde{R}$ *with respect to* $\tilde{\Delta}\colon \widetilde{H} \to \widetilde{H} \otimes \widetilde{H}$,

$$\tilde{\Delta}(\widetilde{Z}_k) = \begin{cases} \widetilde{Z}_k \otimes \widetilde{Z}_k & \textit{for } k \equiv 0 \pmod 2 \,, \\ (\widetilde{Z}_k^{\mathrm{t}} \otimes \widetilde{Z}_k^{\mathrm{t}})^{\mathrm{t}} & \textit{for } k \equiv 1 \pmod 2 \,, \end{cases}$$

so that $\tilde{\Delta}$ *induces a comultiplication* $\Delta\colon H \to H \otimes H$. *Together with the fact that* $\varepsilon(\widetilde{Z}_k \bmod R) = 1$ *this makes* H *a bialgebra and* $\gamma\colon E \to H$ *a bialgebra morphism.*

(b) *The map* $\tilde{\imath}\colon \widetilde{H} \to \widetilde{H}$, *defined by* $\tilde{\imath}(\widetilde{Z}_k) = \widetilde{Z}_{k+1}$ *and* $\tilde{\imath}(fg) = \tilde{\imath}(g)\tilde{\imath}(f)$, *is such that* $\tilde{\imath}(\widetilde{R}) \subset \widetilde{R}$. *Hence it induces a linear map* $i\colon H \to H$ *which is the antipode of* (H, m, Δ).

(c) *For any bialgebra morphism* $\gamma'\colon E \to H'$, *where* H' *is a Hopf algebra, there exists a unique Hopf algebra morphism* $\beta\colon H \to H'$ *such that* $\gamma' = \beta \circ \gamma$.

Proof. (a) Since E is a coalgebra with respect to $\Delta(Z) = Z \otimes Z$, we have

$$\tilde{\Delta}(R_0) \subset \mathbb{K}\langle \widetilde{Z}_0 \rangle \otimes R_0 + R_0 \otimes \mathbb{K}\langle \widetilde{Z}_0 \rangle \subset \widetilde{H} \otimes R_0 + R_0 \otimes \widetilde{H} \,.$$

It follows that the ideal generated by R_0 (and by all R_k, where $k \equiv 0 \pmod 2$) is a coideal. One similarly treats R_k, where $k \equiv 1 \pmod 2$, using $(E, m_E^{\mathrm{op}}, \Delta_E^{\mathrm{op}})$ instead of (E, m, Δ).

To take (8.2) and (8.3) into account, let us consider the following identity: let A, B be two matrices in H with $\Delta(A) = A \otimes A$ and $\Delta(B) = (B^{\mathrm{t}} \otimes B^{\mathrm{t}})^{\mathrm{t}}$. Then

$$\begin{aligned} \Delta(AB)_i^k &= \left((A \otimes A)(B^{\mathrm{t}} \otimes B^{\mathrm{t}})^{\mathrm{t}}\right)_i^k \\ &= \sum_j (A \otimes A)_i^j \, (B^{\mathrm{t}} \otimes B^{\mathrm{t}})_k^j \\ &= \sum_{j,r,s} (a_i^r \otimes a_r^j)(b_s^k \otimes b_j^s) = \sum_{r,s} a_i^r b_s^k \otimes (AB)_r^s \,. \end{aligned}$$

Hence

$$\Delta(AB-I)_i^k = \sum_{r,s} a_i^r b_s^k \otimes ((AB)_r^s - \delta_r^s) + \left(\sum_r a_i^r b_r^k - \delta_i^k\right) \otimes 1\,,$$

and we see that the matrix entries of $AB - I$ generate a coideal. This takes care of the first relations in (8.2) and (8.3). The second ones can be treated similarly. Obviously, the map $\tilde{\varepsilon}\colon \widetilde{Z}_k \to I$ descends to a counit.

(b) Clearly, $\tilde{\imath}(R) \subset \widetilde{R}$. Let $i\colon H \to H$ be the induced linear map. We must check that for each $u \in H$, we have

$$m \circ (i \otimes \mathrm{id}) \circ \Delta(u) = m \circ (\mathrm{id} \otimes i) \circ \Delta(u) = \eta\varepsilon(u) \qquad (8.4)$$

(cf. Section 3.2). From the definition of $\tilde{\Delta}$ and $\tilde{\varepsilon}$ it follows that this is true for all matrix entries u of $Z_k = \widetilde{Z}_k \bmod \widetilde{R}$. Since these entries generate H as a ring, it suffices to check that if (8.4) is true for u and v, then it is true for uv. Indeed,

$$\begin{aligned} m \circ (i \otimes \mathrm{id}) \circ \Delta(u)\Delta(v) &= m \circ (i \otimes \mathrm{id}) \circ \left(\sum_k u'_k \otimes u''_k\right)\left(\sum_l v'_l \otimes v''_l\right) \\ &= m\left(\sum_{k,l} i(v'_l) i(u'_k) \otimes u''_k v''_l\right) \\ &= \sum_l i(v'_l)\left(\sum_k i(u'_k) u''_k\right) v''_l \\ &= \eta\varepsilon(u)\left(\sum_l i(v'_l) v''_l\right) = \eta\varepsilon(u)\eta\varepsilon(v)\,. \end{aligned}$$

(c) Let $\gamma'\colon E \to H'$ be a morphism into a Hopf algebra. Put $Y'_k := i^k_{H'}(\gamma'(Y))$. In view of Theorem 3.4 and Proposition 3.7 the entries of Y'_k satisfy the relations (8.1)–(8.3). Hence, the map $\widetilde{B}\colon \widetilde{Z}_k \to Y'_k$ induces an algebra morphism $\beta\colon H \to H'$ which is clearly compatible with Δ and i. □

8.4 Remarks

(a) This construction has several useful variations. First, one can construct a universal map $\bar{\gamma}\colon E \to \overline{H}$ in the class of Hopf algebras with bijective antipodes. To this end, one has only to adjoin matrices $\widetilde{Z}_k$ and relations (8.1)–(8.3) for all $k \in \mathbb{Z}_+ := \{0, 1, 2, \dots\}$.

Second, one can construct a universal map $\bar{\gamma}_d\colon E \to \overline{H}_d$ in the class of Hopf algebras with antipode i satisfying $i^{2d} = \mathrm{id}$. To do this, one should add

$$\widetilde{Z}_k = \widetilde{Z}_{k+2d} \quad \text{for all } k \in \mathbb{Z}_+$$

to the previous relations.

Third, one can construct "formal quantum groups" corresponding to $H, \overline{H}, \overline{H}_d$. They are completions of these rings with respect to the ideal generated by the matrix entries of $Z_k - I$. Hopf algebras—continuous dual of these objects—are quantum analogues of universal enveloping algebras. Some of these were studied by Drinfeld and Jimbo.

(b) If E is a quadratic algebra, like $\underline{\mathrm{end}}(A)$ or $\underline{\mathrm{e}}(A, g)$, it is nice to have a quadratic version of $H, \overline{H}, \overline{H}_d$ in view of the good properties of QA. Since relations (8.2) and (8.3) are not quadratic, we may add central elements t_k and homogenize (8.2) and (8.3) by setting $\widetilde{Z}_k \widetilde{Z}_{k+1} - t_k^2 I$, etc. Clearly, $\tilde{\Delta}(t_k) = t_k \otimes t_k$ will do. We leave it to the reader to make the universal properties of this construction explicit.

8.5 (Co)Representations

Clearly, $A \xrightarrow{\delta} E \otimes A \xrightarrow{\beta \otimes \mathrm{id}} H \otimes A$ (for $E = \underline{\mathrm{end}}(A)$ or $\underline{\mathrm{e}}(A, g)$) is a corepresentation of the "Hopf envelope" H. Using the universal property of H, one can repeat everything that was said in Section 6.10–6.13 for H instead of E.

In particular, we have the following results:

$$\underline{\mathrm{gl}}(A) := \text{Hopf envelope of } \underline{\mathrm{end}}(A)$$

acts upon

$$\begin{array}{llll} A & \text{via } \delta(x_\bullet) = Z_0 \otimes x_\bullet & (\text{or } Z_k \otimes x_\bullet & \text{if } k \equiv 0 \pmod 2)\,, \\ A^{!\mathrm{op}} & \text{via } \delta(\check{x}_\bullet) = Z_1^{\mathrm{t}} \otimes \mathring{\check{x}}_\bullet & (\text{or } Z_{k+1} \otimes \mathring{\check{x}}_\bullet & \text{if } k \equiv 0 \pmod 2)\,. \end{array}$$

In addition,

$$\underline{\mathrm{gl}}(A, g) := \text{Hopf envelope of } \underline{\mathrm{e}}(A, g)$$

acts upon

$$\begin{array}{ll} A^! & \text{via } \delta(x_\bullet) = Z_0 \otimes x_\bullet\,, \\ A^{\mathrm{op}} & \text{via } \delta(\check{x}_\bullet) = Z_1^{\mathrm{t}} \otimes \check{x}_\bullet\,. \end{array}$$

Chapter 9
Frobenius Algebras and the Quantum Determinant

9.1 Definition

A quadratic algebra A is said to be a *Frobenius algebra* (or *Frobenius quantum space*) of dimension d if

(a) $\dim A_d = 1$, $A_i = 0$ for $i > d$.
(b) For all j, the multiplication map $m\colon A_j \otimes A_{d-j} \to A_d$ is a perfect duality.[1]

The algebra A is called a *quantum Grassmann algebra* if, in addition,
(c) $\dim A_i = \binom{d}{i}$.

9.2 Quantum Determinant

Let E be a bialgebra acting upon a Frobenius space A via

$$\delta\colon A \to E \otimes A\ , \quad \delta(A_1) \subset E \otimes A_1\ .$$

Then $\delta(A_i) \subset E \otimes A_i$, and for $i = d$, we get an element $D = \mathrm{DET}(\delta) \in E$ defined by

$$\delta(a) = \mathrm{DET}(\delta) \otimes a \quad \text{for all } a \in A_d\ .$$

Clearly, D is multiplicative (Proposition 3.9):

$$\delta(D) = D \otimes D\ , \quad \varepsilon(D) = 1\ .$$

[1] For example where this pairing is nonsymmetric for $j = d - j$, see [30]. This asymmetry is the reason why the quantum determinant considered in Example 9.6 might be noncentral.

Y. I. Manin, *Quantum Groups and Noncommutative Geometry*,
CRM Short Courses, https://doi.org/10.1007/978-3-319-97987-8_9

In the language of Section 6.13, D defines the dth quantum symmetric power of the comodule A_1.

9.3 Quantum Cramer and Lagrange Identities

Choose bases $\{X_i^{(j)}\}$ in A_j and $\{Y_k^{(l)}\}$ in A_l so that for a fixed $a \in A_d \setminus \{0\}$ we have

$$X_i^{(j)} Y_k^{(d-j)} = \delta_{ik} a \,. \tag{9.1}$$

Define multiplicative matrices $Z^{(i)}$ and $V^{(l)}$ describing in these bases the coaction $\delta\colon A_k \to E \otimes A_k$

$$\delta\big(X_i^{(j)}\big) = \sum_k z_k^{(i)} \otimes X_k^{(j)} \,, \quad \delta\big(Y_i^{(j)}\big) = \sum_m v_m^{(i)} \otimes Y_m^{(j)} \,. \tag{9.2}$$

Applying δ to (9.1) and taking (9.2) into account, we get

$$Z^{(j)}\big(V^{(d-j)}\big)^{\mathrm{t}} = \mathrm{DET}(\delta) I \,, \tag{9.3}$$

where I is the unit matrix. If A is a Grassmann algebra, $X_i^{(j)}$ and $Y_k^{(d-j)}$ are exterior monomials in a basis of A_1, then (9.3) gives the classical *Cramer identities* for $j = 1$ and *Lagrange identities* for general $j > 1$ (when E is commutative).

9.4 Theorem

Let H be a Hopf algebra acting on a Frobenius algebra A as in Section 9.2, D the determinant of this action. Then

(a) *D is invertible.*
(b) *$H/(D-1)$ is a Hopf algebra, with Δ, i and ε induced by H.*

Proof. (a) Since D is a matrix of a one-dimensional corepresentation, it is invertible in view of Proposition 3.7:

$$i(D) = D^{-1} \,.$$

(b) $D - 1$ generates a coideal in H, since

$$\Delta(D-1) = D \otimes (D-1) + (D-1) \otimes 1 \,.$$

The coideal is i-stable, since

$$i(D-1) = D^{-1} - 1 = -D^{-1}(D-1)\,.$$

Finally, $D - 1 \in \operatorname{Ker}\varepsilon$. □

9.5 General and Special Linear Groups of a Quantum Space

Let A be a quadratic algebra. The Hopf envelopes $\underline{\mathrm{gl}}(A)$ and $\underline{\mathrm{gl}}(A, g)$ of $\underline{\mathrm{end}}(A)$ and $\underline{\mathrm{e}}(A, g)$ constructed in Chapter 8 deserve to be called *general linear groups* of A. If, in addition, A is Frobenius, we can define the *special linear groups*

$$\underline{\mathrm{sl}}(A) = \underline{\mathrm{gl}}(A)/(D-1)\,, \quad \underline{\mathrm{sl}}(A, g) = \underline{\mathrm{gl}}(A, g)/(D-1)\,,$$

where D is the appropriate determinant.

9.6 Example

Let $A = \mathbb{K}\langle x_1, \ldots, x_n\rangle/(x_i^2; x_i x_j + q x_j x_i \mid i < j)$, where $q \in \mathbb{K}^\times$. Clearly, A is n-dimensional Frobenius (in fact, even quantum Grassmannian).

Writing $\delta(x_i) := \sum z_i^k \otimes x_k$ and setting $a = x_1 \cdots x_n \in A_n \setminus \{0\}$, we get the formula for calculating $\mathrm{DET}(\delta)$:

$$\prod_{i=1}^{n}\left(\sum_{k=1}^{n} z_i^k \otimes x_k\right) = \mathrm{DET}(\delta) \otimes \prod_{i=1}^{n} x_i\,,$$

which gives (cf. [56])

$$\mathrm{DET}(\delta) = \sum_{s \in S_n} (-q)^{l(s)} z_1^{s(1)} \cdots z_n^{s(n)}.$$

9.7 Example

For $n \geq 2$, let (cf. Section 7.4)

$$B = \mathbb{K}\langle x_1, \ldots, x_n\rangle/(x_i x_j \text{ for } i \neq j \text{ and } (x_i)^2 - (x_j)^2 \text{ for all } i, j).$$

This is a two-dimensional Frobenius algebra. Taking $a = x_i^2$ we obtain for DET(δ) the formula

$$\mathrm{DET}(\delta) = \sum_{k=1}^{n} \left(z_i^k\right)^2 \quad \text{for each } i = 1, \ldots, n\,.$$

One can similarly obtain two-dimensional Frobenius algebras with quadratic determinant starting from $2d$-dimensional Frobenius algebra A and putting $B := A^{(d)} = \mathbb{K} \oplus A_d \oplus A_{2d}$, see also Gurevich's construction of a Yang–Baxter symmetric algebra which is a two-dimensional Frobenius algebra, see [30]. Such an algebra appears as a "q-antisymmetric algebra" associated with a particular type of Hecke-type YBE. Using the "gluing" procedure from [30] it is possible to construct other solutions of YBE leading to Frobenius algebras of dim ≥ 3. The pairing constructed in [30] is, however, asymmetric, so it yields a slightly generalized Frobenius algebra. In [30], examples are constructed with noncentral quantum determinant D. In several publications, e.g., [31], it is proven that there exists *another* quantum determinant confusingly denoted by the same symbol DET, but constructed via different rules: the above ones correspond to relations in the RTT algebra, whereas the other determinant corresponds to relations in the RE (reflection equations) algebra. This "other DET" is always central.

9.8 Example (Drinfeld [20])

Let $\mathbb{K}$ be the field of fractions of a local ring L. Consider the deformed anticommutation relations

$$x^k x^l + x^l x^k = \sum_{i,j} c_{ij}^{kl} x^i x^j\,, \tag{9.4}$$

where c_{ij}^{kl} belong to the maximal ideal of L and

$$c_{ij}^{kl} = c_{ij}^{lk} = -c_{ji}^{kl}\,.$$

Intuitively, this means that the corresponding algebra is a small deformation of the exterior algebra, at least from the point of view of its relations. Clearly, for $c_{ij}^{kl} = 0$, the relations (9.4) define the usual Grassmann algebra. In [20], Drinfeld calculated conditions on c_{ij}^{kl} which must be satisfied in order for relations (9.4) to define a quantum Grassmann algebra over L (and, a fortiori, over $\mathbb{K}$). Let

$$b_{i_1 i_2 i_3}^{j_1 j_2 j_3} = \sum_j c_{i_1 j}^{j_1 j_2} c_{i_2 i_3}^{j j_3}\,, \quad b = \left(b_{i_1 i_2 i_3}^{j_1 j_2 j_3}\right), \quad a = b\left(1 - \frac{b}{3}\right)^{-1}.$$

Then

$$\text{condition (9.4) determines a quantum Grassmann algebra over } L \iff \text{alt}(i)\,\text{sym}(j)\,a_{i_1 i_2 i_3}^{j_1 j_2 j_3} = 0\,. \qquad (9.5)$$

If (9.5) is true, then A is n-dimensional, and we get DET(δ) in $\underline{\text{end}}(A)$. However, it is impossible to calculate DET(δ) explicitly as in Example 9.6, since we do not know the coefficients $d(s)$ in the expressions

$$x_{s(1)} \cdots x_{s(n)} = d(s) x_1 \cdots x_n \text{ mod } R\,.$$

9.9 Example

$M_q(2)^!$ is a Frobenius quantum space of dimension 4. In fact, it is even Grassmannian. We leave to the reader the task to calculate the corresponding DET.

Chapter 10
Koszul Complexes and the Growth Rate of Quadratic Algebras

10.1 Lemma

For a quadratic algebra A, as in Section 6.1, *let*

$$\xi_A = \xi = \sum_{i=1}^{n} x^i \otimes x_i \in A^! \circ A \subset A^! \otimes A\,.$$

Then $\xi^2 = 0$.

Proof. We have

$$\xi^2 = \sum_{i,k} \tilde{x}^i \tilde{x}^k \otimes \tilde{x}_i \tilde{x}_k \quad (\mathrm{mod}\ R) = \sum_{\lambda} X^{\lambda} \otimes X_{\lambda} \quad (\mathrm{mod}\ R)\,,$$

where $R = R(A^! \circ A)$ whereas (X^{λ}) and (X_{μ}) are arbitrary dual bases of $\tilde{A}_2$. Take $X_{\lambda} = \{r_{\alpha}, s_{\beta}\}$ and $X^{\lambda} = \{s^{\alpha}, r^{\beta}\}$, as in proof of Lemma 6.5. Then

$$\xi^2 = \left(\sum_{\alpha} s^{\alpha} \otimes r_{\alpha} + \sum_{\beta} r^{\beta} \otimes s_{\beta}\right) \quad (\mathrm{mod}\ R) = 0\,. \qquad \square$$

Remarks. (a) Theorem 5.2 gives a conceptual explanation of this lemma. In fact

$$\mathrm{Hom}(\mathbb{K}[\varepsilon] \bullet A, A) = \mathrm{Hom}(\mathbb{K}[\varepsilon], A^! \circ A) \quad \text{and} \quad \mathbb{K}[\varepsilon] \bullet A = A\,.$$

Under this correspondence, id_A goes to $\mathbb{K}[\varepsilon] \to A^! \circ A$ and $\varepsilon \mapsto \xi_A$.

(b) More generally, for any morphism $f: B \to A$ in a QA, we obtain an element $\xi_f \in B^! \circ A$ with $\xi_f^2 = 0$. It corresponds to f under the isomorphism of Theorem 5.2:

Y. I. Manin, *Quantum Groups and Noncommutative Geometry*,
CRM Short Courses, https://doi.org/10.1007/978-3-319-97987-8_10

$$\mathrm{Hom}(\mathbb{K}[\varepsilon] \circ B, A) = \mathrm{Hom}(\mathbb{K}[\varepsilon] \circ B^! = A) .$$

We have

$$\xi_A = \xi_{\mathrm{id}_A} ; \quad \xi_f = (\mathrm{id} \circ f)(\xi_B) .$$

10.2 Koszul Complexes $L^\bullet$

For any ring B and $b \in B$, denote by $l(b)$ (resp. $r(b)$ the left (resp. right) multiplication by b in B (or in a B-module). For a morphism f in QA we can define two complexes

$$L^\cdot(f) = \left(B^! \otimes A, r(\xi_f) \text{ or } l(\xi_f)\right) .$$

Since $\xi_f \in B^!_1 \otimes A_1$, these complexes are direct sums of subcomplexes

$$L^p(f) = \bigoplus_{b-a=p} \left(B^!_b \otimes A_a, r(\xi_f) \text{ or } l(\xi_f)\right) , \quad \text{where } p \in \mathbb{Z} .$$

We write $L^p(A) = L^p(\mathrm{id}_A)$, and set $L^{p,a}(f) = B^!_b \otimes A$ for $b = p + a$.

10.3 Proposition

(a) *Suppose that at least one of the rings $B^!$ and A is finite-dimensional. Then all complexes $L^n(f)$ are finite, and we can define the formal series*

$$\chi_f^L(t) := \sum_n \chi(L^n(f))t^n , \quad \chi(L^n(f)) := \sum_a (-1)^a \dim H^a(L^n(f)) .$$

Let $P_A(t) = \sum \dim A_i \, t^i$. We have

$$P_{B^!}(t) P_A(-t^{-1}) = \chi_f^L(t) . \tag{10.1}$$

(b) *With the same assumptions, let*

$$p = \max\{b \mid B^!_b \neq \{0\}\} \quad \text{and} \quad q = \max\{a \mid A_a \neq \{0\}\}.$$

Then, $H^{a,n} := H^a(L^n(f))$, and we have

$$H^{p,0} = B^!_p , \quad H^{-q,0} = A_q . \tag{10.2}$$

In particular, if $L^{\cdot}(f)$ *is acyclic everywhere except for* (10.2), *for either* $r(\xi_f)$ or $l(\xi_f)$, *then*

$$P_{B^!}(t)P_A(-t^{-1}) = t^p + t^{-q}(-1)^q$$

(*where* t^p (*resp.* t^{-q}) *should be interpreted as* 0 *if* p (*resp.* q) *is not defined*).

Proof. It is quite standard: we have

$$\chi(L^n(f)) = \sum_{b-a=n} (\dim B^!_b)(\dim A_a)(-1)^a .$$

Multiplying by t^n and adding up we get (10.1). The rest is clear. □

10.4 Homological Determinant

From relation (10.2) we deduce that the determinant of a bialgebra E acting upon a Frobenius algebra A is defined by a corepresentation of E in one of the cohomology groups of the complex $L(A)$. In fact, one can define the corepresentation of E on this cohomology, assuming that

$$R(A^!) = R(A^!)^{\mathrm{op}} \quad \text{and} \quad E \text{ is a Hopf algebra} .$$

Then if any one-dimensional cohomology of $L(A)$ happens to exist, it gives rise to a corepresentation of E, which can be called a *homological determinant.*

We prove this for the universal situation $E = \underline{\mathrm{gl}}(A)$. Our construction is motivated by the cohomological treatment of Berezinian in superalgebras (see, e.g., [47]). In fact, this construction gives a quantum Berezinian for a class of "super-Frobenius" quantum supergroups.

10.5 Proposition

Assume that $R(A^!) = R(A^!)^{\mathrm{op}}$. *Then the natural action of* $\underline{\mathrm{gl}}(A)$ *on* A *and* $A^{!\mathrm{op}}$ *defined in Section* 8.5 *induces a corepresentation*

$$H^{\cdot}(L(A)) \to \underline{\mathrm{gl}}(A) \otimes H^{\cdot}(L(A)) .$$

Proof. Clearly, $\underline{\mathrm{gl}}(A)$ acts on $A^! \otimes A$ via

$$A^! \otimes A \xrightarrow{\delta_{A^!} \otimes \delta_A} \underline{\mathrm{gl}}(A) \otimes A^! \otimes \underline{\mathrm{gl}}(A) \otimes A \xrightarrow{(m_{\underline{\mathrm{gl}}(A)} \otimes \mathrm{id}) S_{(23)}} \underline{\mathrm{gl}}(A) \otimes A^! \otimes A .$$

Let us calculate the image of ξ_A in the notation of Section 8.5:

$$\sum_i \check{x}_i \otimes x_i = \xi_A \mapsto \sum_{i,j,k} (Z_1^{\mathrm{t}})_i^j \otimes \check{x}_j \otimes (Z_0)_i^k \otimes x_k$$
$$\mapsto \sum_{i,j,k} (Z_1 - Z_0)_j^k \otimes \check{x}_j \otimes x_k = 1 \otimes \xi_A$$

since $Z_1 Z_0 = I$ in view of (8.2). Therefore we get a morphism of complexes

$$\left(A^! \otimes A, r(\xi) \text{ or } l(\xi)\right) \to \left(\underline{\mathrm{gl}}(A) \otimes A^! \otimes A, r(1 \otimes \xi) \text{ or } l(1 \otimes \xi)\right)$$

which induces our corepresentation.

Remark. I do not know whether one really needs the condition $R(A^!) = R(A^!)^{\mathrm{op}}$. Perhaps, one can omit it by slightly changing our construction.

10.6 Koszul Complexes $K^\bullet$: A Construction

We start with the following simple construction. Suppose we are given a diagram of linear spaces (or objects of an abelian category):

$$\begin{array}{ccccc} E & \subset & F & \subset & G \\ \cup & & \cup & & \cup \\ E' & \subset & F' & \subset & G' \end{array} \tag{10.3}$$

which induces natural maps $E/E' \xrightarrow{\alpha} F/F' \xrightarrow{\beta} G/G'$. Then:

(a) $\beta\alpha = 0 \iff E \subset G'$.
(b) If (a) is true, then $H := \operatorname{Ker}\beta/\operatorname{Im}\alpha \cong (F \cap G')/(E + F')$.

Now let $A^{!*} = \bigoplus_a A_a^{!*}$, where star means linear dualization, and consider the following diagram of the type (10.3):

$$\begin{array}{ccccc} A_{a+1}^{!*} \otimes A_1^{\otimes b-1} & \hookrightarrow & A_a^{!*} \otimes A_1^{\otimes b} & \hookrightarrow & A_{a-1}^{!*} \otimes A_1^{\otimes b+1} \\ \cup & & \cup & & \cup \\ A_{a+1}^{!*} \otimes R_{b-1}(A) & \hookrightarrow & A_a^{!*} \otimes R_b(A) & \hookrightarrow & A_{a-1}^{!*} \otimes R_{b+1}(A)\,. \end{array} \tag{10.4}$$

In order to explain the horizontal inclusion maps, recall that

$$A^!_a = A_1^{*\otimes a} \Big/ \left(\sum_{i=0}^{a-2} A_1^{*\otimes i} \otimes R(A)^{\perp} \otimes A_1^{*\otimes a-2-i} \right) .$$

and hence we can identify

$$A^!_a = \bigcap_{i=0}^{a-2} (A_1^{\otimes i} \otimes R(A) \otimes A_1^{\otimes a-2-i}) \subset A_1^{\otimes a} .$$

Therefore, the subspaces

$$A_a^{!*} \otimes A_1^{\otimes b} = \bigcap_{i=0}^{a-2} (A_1^{\otimes i} \otimes R(A) \otimes A_1^{\otimes a+b-2-i}) \subset A_1^{\otimes a+b}$$

clearly form an increasing filtration of $A_1^{\otimes (a+b)}$ as a decreases. As for the lower line, we similarly have

$$\begin{aligned} A_a^{!*} \otimes R_b(A) &= (A_a^{!*} \otimes A_1^{\otimes b}) \cap (A_1^{\otimes a} \otimes R_b(A)) \\ &= \left(\bigcap_{i=0}^{a-2} A_1^{\otimes i} \otimes R(A) \otimes A_1^{\otimes a+b-2-i} \right) \\ &\qquad \cap \left(\sum_{j=a}^{a+b-2} A_1^{\otimes j} \otimes R(A) \otimes A_1^{\otimes a+b-2-j} \right) . \end{aligned} \tag{10.5}$$

These subspaces also form an increasing filtration as a decreases. Finally, to check that (10.4) generates a complex, we must convince ourselves that

$$A_{a+1}^{!*} \otimes A_1^{\otimes b-1} \subset A_{a-1}^{!*} \otimes R_{b+1}(A) ,$$

i.e., that (uses (10.5))

$$\begin{aligned} &\bigcap_{i=0}^{a-1} A_1^{\otimes i} \otimes R(A) \otimes A_1^{\otimes a+b-2-i} \\ &\subset \left(\bigcap_{i=0}^{a-3} A_1^{\otimes i} \otimes R(A) \otimes A_1^{\otimes a+b-2-i} \right) \cap \left(\sum_{j=a-1}^{a+b-2} A_1^{\otimes j} \otimes R(A) \otimes A_1^{\otimes a+b-2-j} \right). \end{aligned}$$

But this inclusion is clear since $\bigcap_{i=0}^{a-1} \subset \bigcap_{i=0}^{a-3}$ and

$$\bigcap_{i=0}^{a-1} \subset (a-1)\text{th term of} \bigcap = (a-1)\text{th term of} \sum .$$

Finally, since

$$A_a^{!*} \otimes A_1^{\otimes b} \;/\; \left(A_a^{!*} \otimes R_b(A)\right) = A_a^{!*} \otimes A_b \,,$$

we get a new Koszul complex

$$K^{a+b,\bullet}(A): \; \cdots \longrightarrow A_{a+1}^{!*} \otimes A_{b-1} \longrightarrow \underset{\substack{\uparrow \\ b\text{th place}}}{A_a^{!*} \otimes A_b} \longrightarrow A_{a-1}^{!*} \otimes A_{b+1} \longrightarrow \cdots$$

In the previous discussion, we neglected several "border effects", so let us describe the lower complexes explicitly:

$$K^{0,\cdot}: \cdots \longrightarrow 0 \longrightarrow \mathbb{K} \longrightarrow 0 \longrightarrow \cdots \tag{10.6}$$

$$K^{1,\cdot}: \cdots \longrightarrow 0 \longrightarrow \underset{(0)}{A_1^{!*} = A_1} \xrightarrow{\text{id}} \underset{(1)}{A_1} \longrightarrow 0 \longrightarrow \cdots \tag{10.7}$$

$$K^{2,\cdot}: \cdots \longrightarrow 0 \longrightarrow \underset{(0)}{A_2^{!*} = R(A)} \longrightarrow \underset{(1)}{A_1^{!*} \otimes A_1 = A_1^{\otimes 2}}$$

$$\longrightarrow \underset{(2)}{A_0^{!*} \otimes A_2} \longrightarrow 0 \longrightarrow \cdots \tag{10.8}$$

We see that, unlike $L^{a+b,\cdot}$, the complex $K^{a+b,\cdot}$ is always *finite*. If all complexes $K^{p,\cdot}$, with the exception of $p = 0$, are acyclic, A is called a *Koszul algebra*. Many nice properties of Koszul algebras are known, cf. [5, 45, 52].

Writing $\chi_A^K(t) = \sum_n \chi\left(K^{n,\cdot}(A)\right) t^n$ we get, as in Section 10.3, the following statement.

10.7 Proposition

The following identity holds

$$P_{A^!}(t) P_A(-t) = \chi_a^K(t) .$$

In particular, if A is Koszul, then

$$P_{A^!}(t) P_A(-t) = 1 .$$

10.8 Complexes $K^{\bullet}(f)$

Let $f: B \to A$ be a morphism in QA. It induces morphisms

$$\mathrm{id} \otimes f: B^{!*} \otimes B \to B^{!*} \otimes A \quad \text{and} \quad f^{!*} \otimes \mathrm{id}: B^{!*} \otimes A \to A^{!*} \otimes A .$$

Imitating the construction of Section 10.6 one can construct a differential in $B^{!*} \otimes A$, depending on f, in such a way that both these maps become complex morphisms. This complex will be denoted by $K^{\bullet}(f)$. It can be obtained by partial dualization of $L^{\bullet}(f)$.

Suppose that $K^p(f)$ is acyclic for $p > 0$. Looking at (10.7) and (10.8) one sees that, in this case, f is an isomorphism while A and B are Koszul algebras.

Now, let us show how to use the Koszul complexes to estimate the "size" of some quadratic algebras.

For a graded algebra A, we sometimes write $P(A,t)$ instead of $P_A(t)$. For two formal series $f_i(t) = \sum_j a_{ij} t^j$, where $i = 1, 2$, we define

$$(f_1 * f_2)(t) := \sum_j a_{1j} a_{2j} t^j .$$

In this notation, we have the following proposition

10.9 Proposition

(a) *Let A, B be Koszul. Then*

$$P(\underline{\hom}(B, A), t) = \left(P(B, -t) * P(A, -t)^{-1}\right)^{-1} . \tag{10.9}$$

(b) *Let B be Koszul and either A or $B^!$ be finite-dimensional. Then for a morphism $f: B \to A$, we have*

$$\chi_f^L(t) = P(A, -t^{-1}) P(B, t)^{-1} . \tag{10.10}$$

Proof. (a) The class of Koszul algebras is stable with respect to ! and $\circ$ (cf. [5]), and therefore with respect to $\bullet$. Hence, using Proposition 10.7 repeatedly, we find

$$\begin{aligned} P(\underline{\hom}(B, A), t) = P(B^! \bullet A, t) &= P(B \circ A^!, -t)^{-1} \\ &= \left(P(B, -t) * P(A^!, -t)\right)^{-1} \\ &= \left(P(B, -t) * P(A, -t)^{-1}\right)^{-1} . \end{aligned}$$

(b) This follows from Propositions 10.3 and 10.7. □

Remark. Below, in order to calculate the degree of "hidden symmetry" of a quadratic algebra, we use the relation (10.9) for $B = A$. On the other hand, equality (10.10) sometimes allows us to estimate a possible range of nontrivial cohomological representations of $\underline{\text{end}}(A)$ on $H^{\bullet}\big(L(A)\big)$.

10.10 Table

A	$P(A,-t)$	$P(A,-t)^{-1}$	$P(A,-t^{-1})$
$S_n = \mathbb{K}[x_1, \ldots x_n]$	$\frac{1}{(1+t)^n} = \sum_{j\geq 0}(-1)^j\binom{j+n-1}{n-1}t^j$	$(1+t)^n = \sum_{k=0}^{n}\binom{n}{k}t^k$	$\frac{t^n}{(1+t)^n}$
$\Lambda_m = S_m^! = \mathbb{K}\langle \xi_1, \cdots, \xi_m\rangle/([\xi_i, \xi_j]_+)$	$(1-t)^m = \sum_{j=0}^{m}(-1)^j\binom{m}{j}t^j$	$\frac{1}{(1-t)^m} = \sum_{k\geq 0}\binom{k+m-1}{m-1}t^k$	$\frac{(1-t)^m}{(-t)^m}$
$E = \mathbb{K}[x_1, \ldots, x_4]/(q_1, q_2)$ (complete intersection of two quadrics = elliptic curve)	$\frac{(1-t)^2}{(1+t)^2} = 1 + 4\sum_{j=0}^{\infty}(-1)^j\, jt^j$	$\frac{(1+t)^2}{(1-t)^2} = 1 + 4\sum_{k=1}^{\infty} kt^k$	$\frac{(1-t)^2}{(1+t)^2}$

Note in particular that

$$P(E,t) = P(S_2 \otimes \Lambda_2, t) \ .$$

Note also that since E and $E^!$ are infinite-dimensional, we cannot calculate χ^L from the table. Is $L^{\boldsymbol{\cdot}}$ essentially acyclic?

10.11 Calculations

(a) $\dim(\underline{\text{end}}(S_2)_i) = (3^{i+1} - 1)/2$. In fact

$$\frac{1}{(1+t)^2} * (1+t)^2 = 1 - 4t + 3t^2 = (1-t)(1-3t) ,$$
$$P(\underline{\text{end}}(S_2), t) = \frac{3/2}{1-3t} - \frac{1/2}{1-t} .$$

(b) $P(\underline{\text{end}}(E), t) = (1+t)^3/(1 - 13t + 19t^2 + t^3)$. The smallest root of the denominator is $0.08915047\ldots$, so $\dim \underline{\text{end}}(E)_i$ grows like $\text{const} \cdot (11.21690\ldots)^i$. Note that $\dim E_1 = 4$ so that 16^i is the growth rate of the corresponding free matrix algebra.

10.12 Problem

How can one systematically calculate $\dim(\underline{\text{end}}(A, g)_i)$? I do not know of any criterion to ensure that $\text{end}(A, g)$ be Koszul. Of course, in some cases, one can find explicit bases in $\underline{\text{end}}(A, g)_i$. Consider, for example, the exceptional case $q^2 = -1$ for $E = \underline{\text{end}}\left(A_q^{2|0}, g\right)$ of Chapter 2 where the "flat" dependence of q breaks down. We can verify that, in this case, E has a basis consisting of monomials

$$a^\alpha b^\beta (cb)^\gamma c^\delta d^\varepsilon .$$

Counting the number of such monomials, one obtains the order of growth corresponding to that of a five-dimensional commutative variety, while for $q^2 \neq -1$ it is four-dimensional.

It is known that in the theory of Hecke algebras the values $q \in \{\text{roots of unity}\}$ generally play a special role. Perhaps, their exceptional properties would show in the structure of $\underline{\text{end}}\left(A_q^{n|0}\right)$, where $A_q^{n|0}$ is given by

$$x_i x_j = q^{-1} x_j x_i \quad \text{for } i < j .$$

Generally speaking, one would expect that a Hecke algebra is a "quantized Weyl group" for $\underline{\text{gl}}\left(A_q^{n|0}\right)$. Can we make this expectation precise and define quantum Weyl groups in a more general setting?

Chapter 11
Hopf ∗-Algebras and Compact Matrix Pseudogroups

11.1 Hopf ∗-Algebras

Let (H, m, Δ) be a Hopf $\mathbb{C}$-algebra with a bijective antipode i. Drinfeld suggested to define a $*$-structure on H by means of an antilinear map $j : E \to E$ with the following properties:

(a) j is an isomorphism of algebras and an anti-isomorphism of coalgebras;
(b) $j^2 = (ij)^2 = \mathrm{id}$.

There is a method for constructing Hopf $*$-algebras. Let E be a bialgebra over $\mathbb{C}$ generated by the entries of a multiplicative matrix Z such that there is a $\mathbb{C}$-isomorphism $\tau\colon (E, m, \Delta) \to (E, m, \Delta^{\mathrm{op}})$ for which $\tau(Z) = Z^{\mathrm{t}}$. Assume that, in addition, $R(E) \subset E_1^{\otimes 2}$ is a real subspace with respect to the real structure generated by tensor products of entries of Z. Then we can define

$$j_E = \{\tau \text{ followed by complex conjugation of coefficients}\}\,.$$

Clearly, j_E satisfies (a) and $j_E^2 = \mathrm{id}$.

11.2 Proposition

Let $\gamma\colon E \to H$ be a universal map of E into a Hopf $\mathbb{C}$-algebra with bijective antipode i. Then there is a unique $*$-structure j on H such that $j\left(\gamma(Z)\right) = \gamma(Z)^{\mathrm{t}}$.

Proof. Recall the explicit construction of (H, γ) in Section 8.2:

$$H = \mathbb{C}\langle \widetilde{Z}_i \mid i \in \mathbb{Z}\rangle / \widetilde{R}\,,$$

Y. I. Manin, *Quantum Groups and Noncommutative Geometry*,
CRM Short Courses, https://doi.org/10.1007/978-3-319-97987-8_11

where $\widetilde{R}$ is generated by the relations (8.1)–(8.3) for all $k \in \mathbb{Z}$. Define an antilinear algebra morphism $\tilde{\jmath}\colon \mathbb{C}\langle \widetilde{Z} \rangle \to \mathbb{C}\langle \widetilde{Z} \rangle$ by setting

$$\tilde{\jmath}(\widetilde{Z}_i) = \widetilde{Z}^{\mathrm{t}}_{-i}\ .$$

Clearly, $\tilde{\jmath}(\widetilde{R}) = \widetilde{R}$, so that $\tilde{\jmath}$ descends to H. Moreover, the antipode of H is induced by $\tilde{\imath}\colon \widetilde{Z}_k \to \widetilde{Z}_{k+1}$. Therefore, one easily sees that $(\tilde{\imath}\tilde{\jmath})^2 = \mathrm{id}$:

$$\widetilde{Z}_k \xrightarrow{\tilde{\jmath}} \widetilde{Z}^{\mathrm{t}}_{-k} \xrightarrow{\tilde{\imath}} \widetilde{Z}^{\mathrm{t}}_{-k+1} \xrightarrow{\tilde{\jmath}} \widetilde{Z}_{k-1} \xrightarrow{\tilde{\imath}} \widetilde{Z}_k\ .$$

We leave the proof of uniqueness to the reader. □

11.3 Example

This construction is applicable to $E = \underline{\mathrm{e}}(A, g)$ (cf. Section 7.2) if $R(A)$ is a real subspace in $A^{\otimes 2}$ with respect to the basis $x_i \otimes x_j$ in which $g = \sum x_i^2$. This construction applied to $\mathrm{SL}_q(2)$ gives, after a completion, the quantum SU(2) considered, e.g., by Woronowicz [69].

11.4 Compact Matrix Pseudogroups

Woronowicz [68] calls a *compact matrix pseudogroup* the following data $(\mathcal{E}, Z, \Delta)$, where

(a) $\mathcal{E}$ is a C^*-algebra, Z is a square matrix whose entries generate a dense $*$-subalgebra E of $\mathcal{E}$.
(b) $\Delta\colon \mathcal{E} \to \mathcal{E} \otimes \mathcal{E}$ is a comultiplication such that $\Delta(Z) = Z \otimes Z$ and Δ is a morphism of C^*-algebras.
(c) E with the induced structure of bialgebra has a bijective antipode i; Woronowicz denoted it by κ.
(d) $i\big(i(a^*)^*\big) = a$ for all $a \in E$.

Let us discuss connections between this notion and the notion of a Hopf $*$-algebra. If a compact matrix pseudogroup $(\mathcal{E}, Z, \Delta)$ is given, then a $*$-structure on E in Drinfeld's sense can be defined by

$$j : a \mapsto i^{-1}(a^*).$$

In fact, j is antilinear since $*$ is; j is multiplicative since both i^{-1} and $*$ are anti-multiplicative, and j reverses comultiplication since i^{-1} does so while $*$ preserves it. Finally,

$$a \xrightarrow{j} i^{-1}(a^*) \xrightarrow{j} i^{-1}\big(i^{-1}(a^*)^*\big) = i^{-1} \circ i(a) = a \quad \text{(in view of (d))}\,,$$

$$a \xrightarrow{ij} a^* \xrightarrow{ij} a\,.$$

Conversely, let H be a Hopf algebra with $*$-structure j. Put

$$a^* = ij(a) \quad \text{for all } a \in H\,.$$

Clearly, this $*$ has all the algebraic properties of a $*$-involution. However, H is not normed. In [68], S.L. Woronowicz suggests the following construction. Consider the class of all $*$-representations π of H in Hilbert spaces. Define

$$\|a\| = \sup_{\pi} \|\pi(a)\|$$

and consider a completion $H \to \overline{H}$ with respect to this seminorm. If this completion is an injection (as in the case of $H = \mathrm{GL}_q(2)$), we get a compact matrix pseudogroup, provided H is generated by the entries of a multiplicative matrix Z.

We strongly recommend the reader to read [68, 69] and subsequent papers by Woronowicz who has actually laid down the fundamentals of the representation theory of compact matrix pseudogroups, i.e., the "compact forms" of our quantum groups.

Chapter 12
Yang–Baxter Equations

12.1 Yang–Baxter Operator

Let F be a linear space, $R: F \otimes F \to F \otimes F$ an invertible linear map. It is well known that if $R = S_{(12)}: f_1 \otimes f_2 \mapsto f_2 \otimes f_1$, then one can define a representation of the symmetric group S_n on $F^{\otimes n}$ by the following prescription: represent each element $\sigma \in S_n$ as a product of transpositions of neighbors and apply $R_{i,i+1} = S_{(i,i+1)}$ instead of each $(i, i+1)$. Of course, such a decomposition is nonunique but the resulting linear operator does not depend on it.

An arbitrary operator R is called a *Yang–Baxter operator* if it shares this property with $S_{(12)}$, i.e., it defines a series of representations of S_n on $F^{\otimes n}$ by the construction described above.

12.2 Proposition

A given operator R is a Yang–Baxter operator if and only if it satisfies the following Yang–Baxter (or triangle) equations:

$$R_{23}R_{12}R_{23} = R_{12}R_{23}R_{12}: F^{\otimes 3} \to F^{\otimes 3} \tag{12.1}$$

$$R_{12}^2 = \mathrm{id}: F^{\otimes 2} \to F^{\otimes 2} \tag{12.2}$$

where, say, $R_{23}(f_1 \otimes f_2 \otimes f_3) = f_1 \otimes R(f_2 \otimes f_3)$, etc.

Proof. This follows directly from the fact that the Coxeter relations between the transpositions define the symmetric group. □

Y. I. Manin, *Quantum Groups and Noncommutative Geometry*,
CRM Short Courses, https://doi.org/10.1007/978-3-319-97987-8_12

12.3 Examples

(a) Let $F = F_{\bar{0}} \oplus F_{\bar{1}}$, where $\{\bar{0}, \bar{1}\} = \mathbb{Z}/2$. Set $\tilde{f} = \bar{0}$ (resp. $\bar{1}$) if $f \in F_{\bar{0}}$ (resp. $f \in F_{\bar{1}}$). Put

$$R(f \otimes g) = (-1)^{\tilde{f}\tilde{g}} g \otimes f \ .$$

This Yang–Baxter operator "generates" in a very definite sense the main notions of superalgebras (cf. below).

(b) More generally, let $F = \bigoplus_\alpha F_\alpha$ where α runs over a set of indices. Define

$$R(f \otimes g) = a_{\alpha\beta}\, g \otimes f \quad \text{for any } f \in F_\alpha,\ g \in F_\beta, a_{\alpha\beta} \in \mathbb{K} \ .$$

Then (12.1) is automatically satisfied while (12.2) means that

$$a_{\alpha\beta} a_{\beta\alpha} = 1 \quad \text{for all}\, \alpha, \beta \ .$$

(c) Let H be a Hopf algebra and F_1, F_2 be H-comodules given by means of the maps $\delta_i : F_i \to H \otimes F_i$. On $F_1 \otimes F_2$ we can define the structure of an H-comodule by means of the following composed map:

$$F \otimes F_2 \xrightarrow{\delta_1 \otimes \delta_2} H \otimes F_1 \otimes H \otimes F_2 \xrightarrow{S_{(23)}} H \otimes H \otimes F_1 \otimes F_2 \xrightarrow{m \otimes \mathrm{id}} H \otimes F_1 \otimes F_2 \ .$$

We recommend the reader to check the axioms of Section 3.8 and calculate the corresponding multiplicative matrix (cf. Proposition 3.9).

Of course, $F_1 \otimes F_2$ is also a H-comodule, so $F \otimes F$ has *two* natural structures of a comodule. However, these comodules may be nonisomorphic!

If we apply this construction to the "universal" case $F = H$ and ask "when the two natural corepresentations on $H \otimes H$ are equivalent?" we can investigate the case when this equivalence is established by means of a conjugation operator:

$$R \colon H \otimes H \to H \otimes H \ , \qquad R(x) = \rho x \rho^{-1} \quad \text{for some } \rho \in H \otimes H \ .$$

We leave the task of rewriting (12.1) and (12.2) in this context to the reader: compare [45] and [56], cf. also the discussion of the "triangle" Hopf algebras in [21].

12.4 Quadratic Algebras Generated by a Yang–Baxter Operator

Let $R \in \mathrm{End}(F \otimes F)$ be a Yang–Baxter operator. Then we can form an *R-symmetric algebra* of F:

$$S_R^{\bullet}(F) = \bigoplus_{n \geq 0} \{S_n\text{-invariants of } F^{\otimes n}\} ,$$

and an *R-exterior algebra* of F:

$$\bigwedge\nolimits_R^{\bullet}(F) = \bigoplus_{n \geq 0} \{f \in F^{\otimes n} \mid \sigma(f) = \operatorname{sgn}(\sigma)\, f \text{ for any } \sigma \in S_n\} .$$

One can easily check that these algebras are, in fact, quadratic:

$$S_R^{\bullet}(F) \longleftrightarrow \{F, (\mathrm{Id} - R)(F \otimes F)\} ,$$
$$\bigwedge\nolimits_R^{\bullet}(F) \longleftrightarrow \{F, (\mathrm{Id} + R)(F \otimes F)\}$$

in the notation of Chapter 3. For $R = S_{(12)}$, we obtain the usual polynomial and exterior algebras. Note that $S_R^{\bullet}(R)^! = \bigwedge_{R^{\perp}}^{\bullet}(F^*)$.

Starting with such an algebra A, we can generate $\underline{\mathrm{end}}(A)$, $\underline{\mathrm{e}}(A, g)$ and their Hopf envelopes $\underline{\mathrm{gl}}(A)$, $\underline{\mathrm{gl}}(A, g)$.

In particular, for the space with a marked basis

$$F = \bigoplus_{i=1}^{n} \mathbb{K}x_i , \quad R(x_i \otimes x_j) = q^{j-i} x_j \otimes x_i \tag{12.3}$$

we can repeat word-for-word everything we did in Chapter 2. Namely, $S_R^{\bullet}(F)$ defines a "quantized" n-space $A_q^{n|0}$, while $\bigwedge_R^{\bullet}(F)$ defines a similar superspace $A_q^{0|n}$ which is Grassmannian (see Definition 9.1) and hence allows us to define a quantum determinant. In this way we obtain the quantum groups $\mathrm{GL}_q(n)$ and $\mathrm{SL}_q(n)$. Supplying them with a $*$-structure as in Chapter 11 we get Woronowicz's $\mathrm{U_q(x)}$.

Of course, we are not obliged to restrict ourselves to a 1-parameter deformation of $\mathrm{GL}(n)$. As in Example 12.3(b), we can consider a more general Yang–Baxter operator in $F = \bigoplus_{i=1}^{n} \mathbb{K}x_i$:

$$R(x_i \otimes x_j) = a_{ij}\, x_j \otimes x_i , \quad \text{where } a_{ij} a_{ji} = 1 \tag{12.4}$$

and again repeat the construction of Chapter 2.

Warning: If $a_{ii} = 1$ for all i, then $\bigwedge_R^{\bullet}(F)$ is still Grassmannian. However, if $a_{ii} = 1$ for some i and $a_{ii} = -1$ for other i, we get a kind of *intrinsic* $\mathbb{Z}/2$-grading on F and our quantum $\mathrm{GL}_R(F)$ has some resemblance to the general linear supergroup $\mathrm{GL}(r|s)$. In order to understand more clearly what does this mean, let us digress a bit and discuss the role of the transposition operator S_σ, where $\sigma \in S_n$, in our principal constructions.

12.5 Transposition Operators

We invite the reader to turn to Chapter 3 and look at the diagrams defining bialgebras, Hopf algebras and various constructions on them. We see that $S_{(12)}$ explicitly appears in at least the following places:

(a) in the connecting axiom (Section 3.1);
(b) in the definitions of m^{op} and Δ^{op} (Section 3.2);
(c) in Theorem 3.4 on the uniqueness of the antipode.

Besides, in Chapter 3, we omitted checks and proofs referring the reader to Abe [1]. If one looks carefully which properties of $S_{(12)}$ are used in these proofs, one finds that they are precisely the Yang–Baxter equations (12.1) and (12.2)!

Therefore, having chosen a Yang–Baxter operator $R \in \text{End}(F \otimes F)$, one can define a whole series of "R-relativized" notions by just replacing $S_{(12)}$ with R in all definitions and proofs. Here is a list of possibilities for the reader to get acquainted with this idea:

- define a structure of R-Hopf algebra on F;
- translate the main construction of Chapter 4 to quadratic algebras $\{F, R(F)\}$ using R instead $S_{(12)}$;
- construct $\underline{\text{end}}_R(A)$ for an R-quadratic algebra A and prove that it is an R-bialgebra.

In particular, making everything $\mathbb{Z}/2$-graded and using the "sign rule" of Example 12.3(a) one gets a superization of our class of quantum groups (and quantum spaces).

Of course, in general, it is rather awkward to have only one Yang–Baxter operator acting on the space $F \otimes F$. One should be able to transpose different linear spaces, to form tensor products of Yang–Baxter operators, to dualize, etc. There is a very convenient axiomatic framework for all this; namely, that of tensor categories. In the next section we briefly review it.

12.6 Weak Yang–Baxter Operators and Braid Groups

An invertible element $R \in \text{End}(F \otimes F)$ satisfying (12.1), but not necessarily (12.2), is called a *weak Yang–Baxter operator*.

Having a weak Yang–Baxter operator, we can define an action of the Artin braid group Bd_n upon each tensor product $F^{\otimes n}$ in such a way that $R_{(i,i+1)}$ corresponds to a standard generator τ_i of the braid group, say, the ith strand goes over the $(i+1)$th one. In fact, here is a graphic representation of (12.1) in terms of braids:

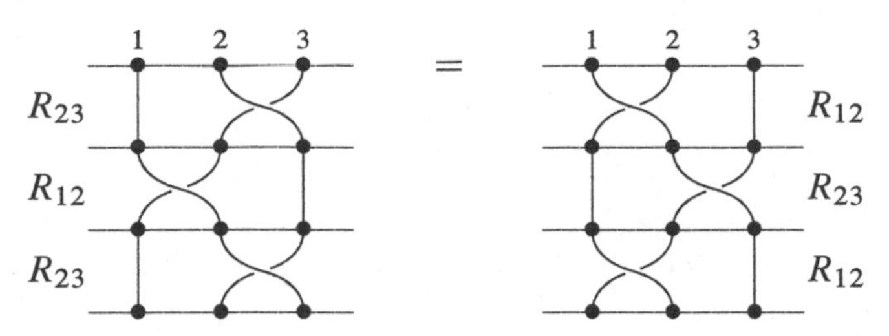

Weak Yang–Baxter operators were recently extensively used in knot theory and related domains: cf. papers by Jones, Turaev, Frlich.

However, I cannot connect these operators directly with quantum groups. Of course, formally one still can repeat the construction of Chapter 5, but the resulting algebras will be probably too small: see Proposition 7.8. The correct way to enlarge them is described in Section 7.9, one should use matrix relations (7.2):

$$R(Z \otimes Z) = (Z \otimes Z)R ,$$

which are only "a part" of the relations defining $\underline{\text{end}}(S^{\cdot}_R(F))$ if $R^2 \neq 1$.

12.7 Summary

Quantum groups are related to Yang–Baxter operator in two ways.

First, a Yang–Baxter operator defines a specific quantum space whose quantum symmetry groups can be constructed in the way described above.

Second, a Yang–Baxter can be implanted in the very definitions of all fundamental objects, thus changing them drastically. The best way to describe this systematically is to start with a collection of, say, linear spaces, stable with respect to tensor product and dualization and endowed with a collection of Yang–Baxter operators $V \otimes W \to W \otimes V$ defined for each pair of this collection of spaces.

In short, we start with an abstract "tensor category" and repeat everything in it. We get a new theory with the same logical structure.

In the next chapter we describe it in more detail.

Chapter 13
Algebras in Tensor Categories and Yang–Baxter Functors

13.1 Tensor Categories: A Summary

In this chapter we reproduce in a very condensed manner the axiomatization of a class of categories which are sufficiently similar to the category of vector spaces over a field $\mathbb{K}$ to allow us to reproduce most of the previous constructions in an abstract setting. See [16] for details and [6] for more references and some important constructions.

Let $\mathcal{S}$ be a category and

$$\otimes : \mathcal{S} \times \mathcal{S} \to \mathcal{S}\ , \quad (X, Y) \mapsto X \otimes Y$$

a bifunctor. An *associativity constraint* for $(\mathcal{S}, \otimes)$ is a functorial isomorphism

$$\varphi_{X,Y,Z} : X \otimes (Y \otimes Z) \xrightarrow{\sim} (X \otimes Y) \otimes Z$$

satisfying a "pentagonal axiom" which implies that one can write a tensor product of an arbitrary finite family of objects without bothering about brackets. A *commutativity constraint* is a functorial isomorphism

$$\psi_{X,Y} : X \otimes Y \xrightarrow{\sim} Y \otimes X$$

satisfying a compatibility condition with φ which implies that for any family of objects $X_1, \dots, X_n$ and $\sigma \in \mathrm{S}_n$ one can define an isomorphism

$$S_\sigma : X_1 \otimes \cdots \otimes X_n \xrightarrow{\sim} X_{\sigma^{-1}(1)} \otimes \cdots \otimes X_{\sigma^{-1}(n)}$$

in such a way that $S_\sigma S_\tau = S_{\sigma\tau}$. Of course, S_σ are composed from elementary transpositions ψ_{X_i, X_j}.

Y. I. Manin, *Quantum Groups and Noncommutative Geometry*,
CRM Short Courses, https://doi.org/10.1007/978-3-319-97987-8_13

An identity object U of $\mathcal{S}$ is a pair $(U, u\colon U \xrightarrow{\sim} U \otimes U)$ such that the functor $\mathcal{S} \to \mathcal{S}\colon X \mapsto U \otimes X$ naturally extends to an equivalence of categories.

A *tensor category* is a category $\mathcal{S}$ together with a tensor product functor $\otimes$, compatible constraints of associativity and commutativity, and a unit object.

Examples. (a) Vector spaces over any field satisfying the conditions

$$\varphi\big(x \otimes (y \otimes z)\big) = (x \otimes y) \otimes z\ , \quad \psi(x \otimes y) = y \otimes x$$
$$\text{for any } x \in X, y \in Y, z \in Z\ .$$

(b) $\mathbb{Z}/2$-graded vector spaces over any field with the same φ and

$$\psi(x \otimes y) = (-1)^{\bar{x}\bar{y}}\, y \otimes x\ , \qquad \text{where } \bar{x} \text{ is the parity of } x\ .$$

(c) The category of quadratic algebras QA with either $\circ$ or $\bullet$ as the tensor product, see Chapter 4.
(d) A category of representations of an affine group scheme (cf. [16]).
(e) A category of (co)representations of a Hopf algebra with an additional structure required to define φ and ψ satisfying the necessary axioms (cf. [45]).

13.2 Rigid Tensor Categories

In a rigid tensor category, for any pair of objects X, Y, one can define their internal $\underline{\mathrm{Hom}}$, i.e., the object $\underline{\mathrm{Hom}}(X, Y)$ together with a functorial isomorphism

$$\underline{\mathrm{Hom}}\big(Z, \underline{\mathrm{Hom}}(X, Y)\big) \xrightarrow{\sim} \underline{\mathrm{Hom}}(Z \otimes X, Y)$$

and functorial isomorphisms

$$\bigotimes_{i \in I} \underline{\mathrm{Hom}}(X_i, Y_i) \xrightarrow{\sim} \underline{\mathrm{Hom}}\Bigg(\bigotimes_{i \in I} X_i, \bigotimes_{i \in I} Y_i\Bigg)$$

(cf. [16]). All examples of the previous section (with appropriate finiteness conditions) are rigid tensor categories, with the exception (QA, $\circ$) where there is no internal $\underline{\mathrm{Hom}}$.

In particular, in any rigid category there is a dualization factor

$$X \to X^{\vee} = \underline{\mathrm{Hom}}(X, U)\ , \quad \text{where } U \text{ is a unit object}\ ,$$

and all objects are reflexive, i.e.,

$$X^{\vee\vee} \cong X\,. \tag{13.1}$$

13.3 Abelian Tensor Category

A rigid tensor category $\mathcal{S}$ is said to be *abelian* if it is an abelian category and $\otimes$ is biadditive. We usually also assume that $\mathrm{Hom}(X, Y)$ are linear spaces over a field $\mathbb{K}$, i.e., that $\mathcal{S}$ is a $\mathbb{K}$-category. The tensor product is then an exact functor.

13.4 Yang–Baxter Functors

Let $(\mathcal{S}, \otimes)$ and $(\mathcal{T}, \otimes)$ be two tensor categories. A faithful functor $F : \mathcal{S} \to \mathcal{T}$ is called a *Yang–Baxter functor* if it transforms the tensor product in $\mathcal{S}$ into the tensor product in $\mathcal{T}$ and is compatible with the associativity constraints in $\mathcal{S}$ and $\mathcal{T}$ but *not necessarily with the commutativity constraints.*

Speaking somewhat loosely, one can imagine an object of $\mathcal{S}$ as an object of $\mathcal{T}$ endowed with an additional structure such that the commutativity constraint in $\mathcal{S}$ differs from that in $\mathcal{T}$ in a way that takes into account this additional structure.

Example. $\mathcal{S}$ = $\mathbb{Z}/2$-graded linear spaces, $\mathcal{T}$ = linear spaces, F = forgetful functor.

Lyubashenko [45] describes how to construct a rigid tensor category $(\mathcal{S}, \otimes)$ with a Yang–Baxter functor to vector spaces starting from *just one* Yang–Baxter operator satisfying a certain nondegeneracy assumption. Essentially one adds all tensor products, duals, subspaces, and quotient spaces "compatible" with the original Yang–Baxter operator so that it generates the new commutativity constraint on the obtained category. (Of course, morphisms are also restricted by a compatibility condition.) Lyubashenko has also proved a theorem showing that such Yang–Baxter functors are essentially forgetful functors from the category of representations of a Hopf algebra.

13.5 Algebra in Abelian Tensor Categories

Now, fix an abelian tensor category $(\mathcal{S}, \otimes)$. Let me explain how one can imitate the standard algebraic constructions in order to define the relativized notions of an $\mathcal{S}$-algebra, $\mathcal{S}$-coalgebra, $\mathcal{S}$-Lie algebra, $\mathcal{S}$-quantum group, etc.

A word of warning is in order. Our axioms are modeled upon the properties of *finite-dimensional* vector spaces. This is essential everywhere when one needs the reflexive properties, see (13.1). On the other hand, a symmetric algebra of a vector space is infinite-dimensional. Therefore we must, in fact, extend our initial universe $(\mathcal{S}, \otimes)$ so as to include infinite direct sums, projective and injective limits, etc. We pretend that all this has somehow been done and forget about it (for a version of the solution of this important problem, see [6]).

13.5.1 $\mathcal{S}$-*Algebra*

An *associative* $\mathcal{S}$-*algebra* is an object A in $\mathcal{S}$ together with a multiplication morphism $m\colon A \otimes A \to A$ satisfying the associativity axiom of Section 3.1. Similarly, an associative $\mathcal{S}$-algebra with a unit consists of A, m, and a morphism $\eta\colon U \to A$ satisfying the axiom of unit of 3.1. Here U is a unit object of $\mathcal{S}$.

It is clear now how to define an $\mathcal{S}$-algebra, $\mathcal{S}$-bialgebra, $\mathcal{S}$-Hopf algebra imitating Chapter 3. The notion of an $\mathcal{S}$-Lie algebra is probably less evident. It is convenient to start with the definition of $\mathcal{S}$-commutator.

13.5.2 $\mathcal{S}$-*Commutator*

Let (A, m) be an associative $\mathcal{S}$-algebra. Let us return to the S_σ-notation instead of $\psi_{X,Y}$-notation and define the $\mathcal{S}$-commutator map:

$$[\cdot]_{\mathcal{S}} : A \otimes A \to A\,, \quad [f]_{\mathcal{S}} = m(f) - mS_{(12)}(f)\,.$$

For $\mathcal{S} = \{\text{vector spaces}\}$ we get the usual formula

$$[a \otimes b]_{\mathcal{S}} = ab - ba\,.$$

One easily checks the following identities

$$[S_{(12)}(f)]_{\mathcal{S}} = -[f]_{\mathcal{S}} \qquad \text{“}\mathcal{S}\text{-anticommutativity”} \qquad (13.2)$$

$$\{g\}_{\mathcal{S}} + \{S_{(312)}(g)\}_{\mathcal{S}} + \{S_{(231)}(g)\}_{\mathcal{S}} = 0\,, \qquad \text{“}\mathcal{S}\text{-Jacobi identity”} \qquad (13.3)$$

where

$$\{\cdot\}_{\mathcal{S}} : A \otimes A \otimes A \xrightarrow{\mathrm{id}\otimes[\cdot]_{\mathcal{S}}} A \otimes A \xrightarrow{[\cdot]_{\mathcal{S}}} A \qquad (13.4)$$

is the triple commutator $\big[a, [b, c]_{\mathcal{S}}\big]_{\mathcal{S}}$.

We can now define $\mathcal{S}$-*commutative algebras* as those for which $[\cdot]_{\mathcal{S}} = 0$ identically. Rewriting this condition in terms of a commutative diagram, we can also reverse arrows and define $\mathcal{S}$-*cocommutative* $\mathcal{S}$-*algebras*, etc. It is instructive to look at the place where the definitions of $A^!_{\mathcal{S}}$, $A^{\mathrm{op}}_{\mathcal{S}}$, $A \otimes_{\mathcal{S}} B$, etc. differ from those of $A^!$, $[2]A^{\mathrm{op}}$, $A \otimes B$, etc.

13.5.3 $\mathcal{S}$-*Lie Algebra*

An $\mathcal{S}$-*Lie algebra* is an object L of $\mathcal{S}$ together with a morphism $[\cdot]_{\mathcal{S}} : L \otimes L \to L$ satisfying (13.2) and (13.3) (note that the triple commutator (13.4) is defined in terms of $[\cdot]_{\mathcal{S}}$ only).

13.5.4 *Differential Operators*

Let A be an associative $\mathcal{S}$-algebra. Following Grothendieck, we can define an internal object of $\mathcal{S}$-differential operators $\mathrm{Diff}_{\leq i}(A)$ of order $\leq i$ together with the left action map

$$l_i : \big(\mathrm{Diff}_{\leq i}(A)\big) \otimes A \to A$$

by the following inductive procedure. Set

$$\mathrm{Diff}_0(A) = A\ , \quad l_0 = m\ .$$

Furthermore, $\mathrm{Diff}_{i+1}(A)$ is the maximal subobject of $\underline{\mathrm{Hom}}(A, A)$ (which is an $\mathcal{S}$-algebra) such that

$$[\mathrm{Diff}_{\leq i+1}(A), \mathrm{Diff}_0(A)]_{\mathcal{S}} \subset \mathrm{Diff}_{\leq i}(A)$$

and l_{i+1} is induced by the canonical map $\mathrm{Hom}(A, A) \otimes A \to A$. Of course, the existence of such a subobject is a separate matter. Having $\mathrm{Diff}_{\leq i}$, one can formally introduce quantum jet spaces.

13.5.5 *Quadratic* $\mathcal{S}$-*Algebras*

The constructions of Chapters 4 and 5 can be repeated word for word.

We hope that the attentive reader has by now grasped the rules of the game. One of the interesting outcomes is a rich new supply of complexes defining $\mathcal{S}$-relativized versions of Koszul algebras, Hochschild (co)homology, cyclic (co)homology, etc.

13.6 A Proposal for a Noncommutative Algebraic Geometry Universe

Take an abelian tensor $\mathbb{K}$-category $(\mathcal{S}, \otimes)$ such that there exists a Yang–Baxter functor $(\mathcal{S}, \otimes) \to \{\mathbb{K}\text{-vector spaces}\}$. Consider the category of $\mathcal{S}$-commutative graded algebras A and try to initiate the Serre-style projective algebraic geometry in this category. In particular such an algebra A defines, in an intrinsic way, the abelian category "Coh Proj A" of "coherent sheaves on Proj A": just factorize the category of *graded* $\mathcal{S}$–A-modules by *finite* ones. Morphisms between such projective spectra should be defined as certain functors among "Coh Proj A" ("inverse images"). A realization of this program has been started by A. Verevkin [65, 66].

In short, as A. Grothendieck taught us, to do geometry you really don't need a space, all you need is a category of sheaves on this would-be space, see

http://www.grothendieckcircle.org.

Chapter 14
Some Open Problems

In conclusion we state here in a rather unsystematic way some questions and indicate directions for further research related to the approach to quantum groups described in these lectures.

14.1 Differential Geometry

One should be able to define the notions of the de Rham complex, tangent and cotangent spaces, etc., for a quantum quadratic space and its end-bialgebra and then transport them to the corresponding quantum group. There is no doubt that it can be done in a standard way in the category of quadratic $\mathcal{S}$-commutative algebras, where $\mathcal{S}$ is a Yang–Baxter category of linear spaces. The Woronowicz differential forms (see [69]) on SU(2) seem to fit in this framework.

Is it possible to significantly broaden the scope of this differential geometry? The Koszul complex $K^*(A)$ vaguely looks like a complex of noncommutative currents (not forms, due to an additional dualization $A^! \to A^{!*}$).

Of course, we have Alain Connes' suggestion: to use cyclic cohomology.

14.2 Cyclic (Co)homology

For quadratic algebras A this was studied by Feigin and Tsygan. One should understand whether the "quantum symmetries" of A act on its Hochschild and cyclic (co)homology and study the cyclic (co)homology of quantum (semi)groups.

Y. I. Manin, *Quantum Groups and Noncommutative Geometry*,
CRM Short Courses, https://doi.org/10.1007/978-3-319-97987-8_14

14.3 Root Technique and Kac–Moody Quantum Groups

Chevalley, Springer, and others have applied the classical root technique directly to the ring of functions on an algebraic group instead of its Lie algebra. Can one extend this approach to our quantum groups, perhaps by first quantizing the simplest fundamental representation treated as a "quadratic quantum space"? Can one then use some (version of) Kac–Moody Cartan matrices?

14.4 Quantum Virasoro?

The same question for the Virasoro "group." One possible approach is to use the embedding of the Virasoro algebra into the (central extension of) an infinite-dimensional linear Lie algebra. For such embeddings, see Neretin's paper [50]. For Lie superalgebras generalizing the Virasoro central extension of simple Lie algebra, see [29, 38]. Since we have a rich supply of quantum linear groups and some of our constructions modify easily to define various kinds of $GL_{\text{quant}}(\infty)$, we may hope also to quantize Virasoro. The new effects of infinite dimension should be studied carefully. And cautiously: on various versions of $GL(\infty)$ at the level of Lie superalgebras, see [23].

14.5 Flag Spaces, Quantum Fibrations, Yang–Mills

It would be very important to define noncommutative flag spaces for quantum groups or, at least, principal bundles over them. This chapter of noncommutative (algebraic) geometry is wide open. Of course, one hopes that for "good" groups (like the symmetries of $\mathcal{S}$-symmetric algebras) one should obtain Schubert cells, Borel–Weil–Bott theory, etc. Here, one should not act too hastily since even in supergeometry this program was started only recently and revealed both rich content and some puzzling new phenomena, see [47].

14.6 Representations of Rings of Functions

The initial source of quantum group theory, the quantum inverse transform method requires the study of (unitary) representations of function rings of quantum groups.

See also articles by Woronowicz [68, 69] and Vaksman–Soibelman [61, 64].

14.7 Koszul Rings and Differential Graded Algebras

The class of quadratic Koszul algebras is a very important one. Among its many interesting properties one should mention that if A is Koszul, then $A^!$ is canonically isomorphic to $\mathrm{Ext}^{\bullet}_A(\mathbb{K}, \mathbb{K})$. (Otherwise it is a superalgebra generated by Ext^1.) This makes one wonder whether it is possible to generalize our constructions of quadratic algebras extending them to a certain category of differential graded algebras (with morphisms considered up to homotopy).

14.8 "Hidden Symmetry" in Algebraic Geometry

An arbitrary projective algebraic variety (actually, even a scheme) is a projective spectrum of a quadratic algebra. For some very important varieties like abelian varieties (with rigidity) and their moduli spaces, such quadratic algebras come equipped with canonical generators of degree one (Mumford's theory of abstract theta-functions). Our results show that certain universal Hopf algebras act on such manifolds. Their properties deserve closer investigations.

Probably, a simpler class of examples is furnished by canonical embeddings of algebraic curves. With rare exceptions, a curve of genus g embeds into $\mathbb{P}^{g-1}$ with the help of its differentials of the first kind, and all relations among them are generated by quadratic ones. The corresponding Hopf algebras of hidden symmetries of a curve may, in principle, be much larger than the rings of functions on classical automorphism groups (which are finite-dimensional for $g \geq 2$ and usually trivial).

Chapter 15
The Tannaka–Krein Formalism and (Re)Presentations of Universal Quantum Groups

Theo Raedschelders and Michel Van den Bergh

15.1 Introduction

In this chapter we expand upon Section 14.8 of the present volume, where the possibility of "hidden symmetry" in algebraic geometry was discussed. More precisely, our goal is to convince the reader that, as long as one starts with a reasonable algebra A, the universal bi- and Hopf algebras $\underline{\mathrm{end}}(A)$ and $\underline{\mathrm{gl}}(A)$ introduced in Chapters 5 and 8 are well-behaved objects.

To do so, it seems natural to start by looking at $A = \mathbb{K}[x_1, \ldots, x_n]$, which we think of as a homogeneous coordinate ring for projective space. Our work in [53–55] has focused on the representation theory of the universal Hopf algebra $\underline{\mathrm{gl}}(A)$ coacting on A (and on its noncommutative counterparts). We show the representations are as nice as can be: the category of comodules for $\underline{\mathrm{gl}}(A)$ can be given the structure of a highest weight category (see Definition 15.28), and it shares many more similarities with the category of rational representations of the general linear group GL_n, or equivalently, the category of comodules over the coordinate Hopf algebra $\mathcal{O}(\mathrm{GL}_n)$. In contrast to $\mathcal{O}(\mathrm{GL}_n)$, however, the universal Hopf algebras $\underline{\mathrm{gl}}(A)$ have rather complicated presentations and are, moreover, of exponential growth. In order to deal with them, we resort to using a different set of techniques, which go by the name of the Tannaka–Krein formalism.

Y. I. Manin, *Quantum Groups and Noncommutative Geometry*,
CRM Short Courses, https://doi.org/10.1007/978-3-319-97987-8_15

15.2 The Tannaka–Krein formalism

15.2.1 *The Basic Example*

Consider a finite group G. The starting point for the Tannaka–Krein formalism is the basic question: can G be recovered from its category of finite-dimensional complex representations $\mathrm{rep}_{\mathbb{C}}(G)$?

As stated, this question is at best unclear, since one needs to specify what structure on the category $\mathrm{rep}_{\mathbb{C}}(G)$ is taken into account. Indeed, one can consider $\mathrm{rep}_{\mathbb{C}}(G)$ as:

(1) (abelian) category,
(2) monoidal category,
(3) symmetric monoidal category.

For (1), the answer is no: for any finite group H with the same number of conjugacy classes as G, say n, there is an equivalence of categories $\mathrm{rep}_{\mathbb{C}}(G) \cong \mathrm{rep}_{\mathbb{C}}(H)$, since by the Artin–Wedderburn theorem both categories are equivalent to the category $\mathrm{mod}(\mathbb{C}^n)$.

For (2), the answer is also no, though this is more subtle. Consider, for example, the two nonabelian groups of order 8, the dihedral group D_8 and the quaternion group Q_8. These groups have the same character table, and even isomorphic Grothendieck rings, but one can check that $\mathrm{rep}_{\mathbb{C}}(\mathrm{D}_8)$ and $\mathrm{rep}_{\mathbb{C}}(\mathrm{Q}_8)$ are not equivalent as monoidal categories. Nevertheless, there are two nonisomorphic groups G and H of order 64, both of which arise as semidirect products of $\mathbb{Z}/2 \times \mathbb{Z}/2$ and $\mathbb{Z}/4 \times \mathbb{Z}/4$, such that $\mathrm{rep}_{\mathbb{C}}(G) \cong \mathrm{rep}_{\mathbb{C}}(H)$ as monoidal categories, see [35]. Finite groups with monoidally equivalent categories of finite-dimensional complex representations are called *isocategorical* in [25]. In loc. cit. all groups isocategorical to a given group are classified in terms of group-theoretical data.

For (3), the answer is yes: by [17, Theorem 3.2(b)], the forgetful functor

$$F\colon \mathrm{rep}_{\mathbb{C}}(G) \to \mathrm{vect}_{\mathbb{C}} \tag{15.1}$$

is the unique $\mathbb{C}$-linear exact faithful symmetric monoidal functor from $\mathrm{rep}_{\mathbb{C}}(G)$ to $\mathrm{vect}_{\mathbb{C}}$, where $\mathrm{vect}_{\mathbb{C}}$ is the category of finite-dimensional $\mathbb{C}$-vector spaces. Hence, we can assume F is known, and the following proposition then allows one to reconstruct G.

15.1 Proposition. *There is an isomorphism of groups*

$$G \to \mathrm{Aut}^{\otimes}(F)\,,$$

where $\mathrm{Aut}^{\otimes}(F)$ *denotes the group of natural isomorphisms of* F *which are compatible with the tensor product on* $\mathrm{rep}_{\mathbb{C}}(G)$.

Proof. Every element of G acts via a linear map in every finite-dimensional G-representation, so define

$$\phi\colon G \to \mathrm{Aut}^{\otimes}(F)\colon g \mapsto \big(\rho(g)\big)_{(V,\rho)},$$

where $\rho\colon G \to \mathrm{GL}(V)$ is a G-representation. One easily checks that ϕ is a well-defined group morphism. Now assume $\phi(g) = \mathrm{id}_F$, then, in particular,

$$\rho(g) = \mathrm{id}\colon \mathcal{O}(G) \to \mathcal{O}(G)\colon \delta_h \mapsto \delta_{gh},$$

where $(\mathcal{O}(G), \rho)$ is the representation of G on the algebra of functions on G, and δ_h denotes the indicator function at h. This implies that $g = 1$, so ϕ is injective.

Surjectivity is harder to check. For a given $\alpha \in \mathrm{Aut}^{\otimes}(F)$, we again look at the representation $(\mathcal{O}(G), \rho)$. In fact, one can show that $\alpha_{(\mathcal{O}(G),\rho)}$ is an algebra morphism, and from there one shows that there is a unique $g \in G$ such that

$$\alpha_{(\mathcal{O}(G),\rho)}\colon \mathcal{O}(G) \to \mathcal{O}(G)\colon f \mapsto f(-\cdot g).$$

In other words, $\alpha_{(\mathcal{O}(G),\rho)} = \phi(g)_{(\mathcal{O}(G),\rho)}$. This then suffices to ensure that $\alpha = \phi(g)$. We omit the computational details. □

Already from this basic example, it is clear that the algebra of functions $\mathcal{O}(G)$ of G plays a crucial role.

15.2.2 *Tannaka–Krein Reconstruction*

From now on we will work over an arbitrary algebraically closed field $\mathbb{K}$.

The key ingredient in the reconstruction for finite groups was the existence of the forgetful functor F (15.1). In this section we consider the more general setting of a covariant functor $F\colon \mathcal{A} \to \mathrm{vect}_{\mathbb{K}}$, where $\mathcal{A}$ is an

(1) essentially small,
(2) $\mathbb{K}$-linear,
(3) abelian

category. To this data we will associate a certain coalgebra serving as a substitute for $\mathcal{O}(G)$ which occurs in the proof of Proposition 15.1. We will, however, start by taking a more intuitive dual viewpoint.

We first give a brief reminder on pseudo-compact algebras, for more details see [7, Section 4]. For us a *pseudo-compact* algebra A is a topological algebra whose topology is generated by two-sided ideals of finite codimension and which is moreover complete. Denote the category of pseudo-compact algebras with continuous algebra morphisms by $\mathrm{PC}_{\mathbb{K}}$.

Similarly, a right linear topological A-module M is called a *pseudo-compact A-module* if its topology is generated by right A-submodules of finite codimension and M is complete. The corresponding category is denoted $\mathrm{PC}(A)$. It is an abelian category. We will also need the category $\mathrm{Dis}(A)$ of *discrete A-modules*, i.e., the right linear topological A-modules equipped with the discrete topology. These categories are dual in the following sense. There are functors

$$\mathrm{Dis}(A) \underset{(-)^\circ}{\overset{(-)^*}{\rightleftarrows}} \mathrm{PC}(A^{\mathrm{op}})$$

where $(-)^*$ denotes taking the vector space dual and $(-)^\circ$ the continuous dual.[1] These functors define mutually inverse anti-equivalences of categories.

The pseudo-compact algebra associated to $(\mathcal{A}, F)$ is denoted $\mathrm{End}(F)$, and consists of all natural transformations $F \Rightarrow F$, with $\mathbb{K}$-linear structure coming from $\mathrm{vect}_{\mathbb{K}}$, and multiplication defined via composition of natural transformations. The topology is determined by associating to every finite $\alpha \subset \mathrm{Ob}(\mathcal{A})$ a base open set

$$U(\alpha) = \bigcap_{X \in \alpha} \mathrm{Ker}\big(\mathrm{End}(F) \to \mathrm{End}(FX)\big) ,$$

which is an ideal of finite codimension.

It is clear that for every object $X \in \mathcal{A}$, FX is a finite-dimensional discrete $\mathrm{End}(F)$-module, so we can consider the evaluation functor

$$\mathrm{ev}_F : \mathcal{A} \to \mathrm{dis}\big(\mathrm{End}(F)\big) : X \mapsto FX , \tag{15.2}$$

where $\mathrm{dis}\big(\mathrm{End}(F)\big)$ denotes the category of finite-dimensional discrete $\mathrm{End}(F)$-modules. The following theorem is the quintessential *Tannaka–Krein reconstruction theorem.*

15.2 Theorem ([28, 63]). *If F is faithful and exact, then*

$$\mathrm{ev}_F : \mathcal{A} \to \mathrm{dis}\big(\mathrm{End}(F)\big)$$

[1] For a pseudo-compact A-module M, the module M° consists of all the continuous linear functionals $M \to \mathbb{K}$, where $\mathbb{K}$ has the discrete topology.

defines an equivalence of categories.

Proof. If $\mathcal{B}$ is an essentially small abelian category, then the category $\mathrm{Ind}(\mathcal{B})$ of *ind-objects* of $\mathcal{B}$ is the category of left exact contravariant functors $\mathcal{B} \to \mathrm{Ab}$. By [28, Chapter II] $\mathrm{Ind}(\mathcal{B})$ is a Grothendieck category, the Yoneda embedding

$$\mathcal{B} \to \mathrm{Ind}(\mathcal{B}) \colon B \mapsto \mathcal{B}(-, B)$$

is fully faithful and its essential image yields a family of finitely presented generators for $\mathrm{Ind}(\mathcal{B})$. Moreover, if every object in $\mathcal{B}$ is Noetherian, then the essential image of $\mathcal{B}$ coincides with the category of Noetherian objects in $\mathrm{Ind}(\mathcal{B})$.

We will apply this with $\mathcal{B} = \mathcal{A}^{\mathrm{op}}$. Since F is exact and faithful, it follows that $\mathcal{A}$ has finite-dimensional Hom-spaces and every object is of finite length. Therefore $\mathcal{A}^{\mathrm{op}}$ enjoys the same properties. Hence, the Noetherian objects in the Grothendieck category $\mathrm{Ind}(\mathcal{A}^{\mathrm{op}})$ are given by $\mathcal{A}^{\mathrm{op}}$. Since finite length objects are Noetherian, this is then also true for the category of finite length objects.

We claim that F is in fact an injective cogenerator for $\mathrm{Ind}(\mathcal{A}^{\mathrm{op}})$. We first note that $\mathrm{Hom}(-, F)$ coincides with F when restricted to $\mathcal{A}^{\mathrm{op}}$. Indeed the composition is given by

$$A \mapsto \mathcal{A}(A, -) \mapsto \mathrm{Ind}(\mathcal{A}^{\mathrm{op}})(\mathcal{A}(A, -), F) = F(A). \tag{15.3}$$

We see in particular that $\mathrm{Hom}(-, F)$ is exact when restricted to $\mathcal{A}^{\mathrm{op}}$ and therefore F is at least fp-injective, i.e., $\mathrm{Ext}^1(X, F) = 0$ for every finitely presented functor X. Since $\mathrm{Ind}(\mathcal{A}^{\mathrm{op}})$ is locally Noetherian, this implies that F is injective (by [41, A.4, Proposition A.11]). Faithfulness then ensures that F is a cogenerator.

By [28, Chapter IV, Proposition 13] $\mathrm{End}(F)$ is pseudo-compact. Note that the definition in [28, Chapter IV, Section 3] of pseudo-compactness is more general than ours, but using the fact that F takes values in finite-dimensional vector spaces, one checks that $\mathrm{End}(F)$ is indeed pseudo-compact in our sense. A similar statement holds for pseudo-compact $\mathrm{End}(F)$-modules.

By [28, Chapter IV, Theorem 4], there is a commuting diagram

$$\begin{array}{ccc} \mathrm{Ind}(\mathcal{A}^{\mathrm{op}}) & \xrightarrow{\mathrm{Hom}(-,F)} & \mathrm{PC}\big(\mathrm{End}(F)\big) \\ {\scriptstyle \mathrm{Yoneda}}\uparrow & & \uparrow \\ \mathcal{A}^{\mathrm{op}} & \xrightarrow{\mathrm{Hom}(-,F)} & \mathrm{pc}\big(\mathrm{End}(F)\big) \end{array} \tag{15.4}$$

where the horizontal arrows are anti-equivalences of categories, and $\mathrm{pc}\big(\mathrm{End}(F)\big)$ denotes the category of pseudo-compact $\mathrm{End}(F)$-modules of finite length. In particular the lower row gives an equivalence

$$\mathcal{A} \xrightarrow{\operatorname{Hom}(-,F)} \operatorname{pc}\big(\operatorname{End}(F)\big) \tag{15.5}$$

Now we observe that we have, in fact, $\operatorname{pc}\big(\operatorname{End}(F)\big) = \operatorname{dis}(\operatorname{End}(F)$ since the topology on any pseudo-compact $\operatorname{End}(F)$-module is generated by submodules of finite codimension. Combining this observation with (15.3) (which shows that $\operatorname{Hom}(-, F)$ restricted to $\mathcal{A}$ is ev_F) we see that (15.5) ultimately translates into an equivalence

$$\operatorname{ev}_F : \mathcal{A} \to \operatorname{dis}\big(\operatorname{End}(F)\big) . \qquad \square$$

We now introduce the coalgebra which will play the role of $\mathcal{O}(G)$ for the functor $F : \mathcal{A} \to \operatorname{vect}_{\mathbb{K}}$. Using the mutually inverse dualities of categories

$$\operatorname{Coalg}_{\mathbb{K}} \underset{(-)^{\circ}}{\overset{(-)^*}{\rightleftarrows}} \operatorname{PC}_{\mathbb{K}}$$

where $\operatorname{Coalg}_{\mathbb{K}}$ is the category of $\mathbb{K}$-coalgebras, we define

$$\operatorname{coend}(F) := \operatorname{End}(F)^{\circ} .$$

This gives rise to the equivalence of categories

$$\operatorname{comod}\big(\operatorname{coend}(F)\big) \cong \operatorname{dis}\big(\operatorname{End}(F)\big) ,$$

where comod denotes the category of finite-dimensional comodules.

A more concrete description of $\operatorname{coend}(F)$ can be given as follows:

$$\operatorname{coend}(F) = \frac{\bigoplus_{X \in \mathcal{A}} FX^* \otimes FX}{E} , \tag{15.6}$$

where E is the following subspace:

$$E = \langle y_* \otimes (Ff)(x) - (Ff)^*(y_*) \otimes x \mid f \in \operatorname{Hom}(X, Y), x \in FX, y_* \in FY^* \rangle_{\mathbb{K}} ,$$

with comultiplication and counit

$$\Delta([\xi \otimes x]) = \sum_i [\xi \otimes x_i] \otimes [\xi_i \otimes x] ,$$

$$\epsilon([\xi \otimes x]) = \sum_i \xi(x_i)\xi_i(x) ,$$

for $\xi \otimes x \in FX^* \otimes FX$ and $\sum_i \xi_i \otimes x_i \in FX^* \otimes FX$ a dual basis. One checks that Δ and ϵ are well-defined and satisfy the coassociativity and counitality conditions.

15.3 *Remark.* More abstractly, for any functor $G\colon \mathcal{C}^{\mathrm{op}} \times \mathcal{C} \to \mathcal{D}$, where $\mathcal{C}$ is small and $\mathcal{D}$ is cocomplete (i.e., it has all small colimits), the coend(G) can be defined as the colimit

$$\coprod_{c \to c'} G(c', c) \rightrightarrows \coprod_{c} G(c, c) \to \operatorname{coend}(G)\,. \tag{15.7}$$

In our case, one takes $G\colon \mathcal{A}^{\mathrm{op}} \times \mathcal{A} \to \mathrm{Vect}_{\mathbb{K}}\colon X \mapsto FX^* \otimes FX$, where $\mathrm{Vect}_{\mathbb{K}}$ denotes the category of all $\mathbb{K}$-vector spaces.[2]

From now on, we will only work with coend(F) and comod$\big(\mathrm{coend}(F)\big)$. Theorem 15.2 affords a dictionary between categorical structures on the pair $(\mathcal{A}, F)$ and algebraic structures on the coalgebra coend(F). An example of how this dictionary works is provided by the following proposition.

15.4 Proposition. (1) *If $\mathcal{A}$ is monoidal, and if F is a monoidal functor, then* coend(F) *can be made into a bialgebra.*
(2) *If $\mathcal{A}$ moreover has left duals, then* coend(F) *can be made into a Hopf algebra.*
(3) *If $\mathcal{A}$ moreover has right duals (i.e., $\mathcal{A}$ is rigid monoidal), then* coend(F) *can be made into a Hopf algebra with invertible antipode.*

Proof. Consider (2) for example. The antipode is defined as

$$S\colon \operatorname{End}(F) \to \operatorname{End}(F)\colon \phi \mapsto S(\phi),$$

where $S(\phi)_X = \phi^*_{X^*}\colon FX \to FX$ for any $X \in \mathcal{A}$. One then checks that S is continuous and the axioms for left duals in $\mathcal{A}$ correspond exactly to the antipode axiom. □

15.5 *Example.* If G is an affine algebraic group scheme over $\mathbb{K}$, denote by $\mathrm{rep}_{\mathbb{K}}(G)$ the rigid monoidal category of finite-dimensional rational G-representations and $F\colon \mathrm{rep}_{\mathbb{K}}(G) \to \mathrm{vect}_{\mathbb{K}}$ the forgetful (monoidal) functor. Then one computes that

$$\operatorname{coend}(F) \cong \mathcal{O}(G)$$

as Hopf algebras, and ev_F is the familiar equivalence between $\mathrm{rep}_{\mathbb{K}}(G)$ and the category of finite-dimensional comodules over the coordinate ring of G.

[2] The appearance of $\mathrm{Vect}_{\mathbb{K}}$ instead of $\mathrm{vect}_{\mathbb{K}}$ here is to make sure the colimit (15.7) makes sense.

15.3 Presentations of $\underline{\mathrm{gl}}(A)$

The starting point of our work in [54, 55] is the following simple observation: in order to construct the bialgebra (resp. Hopf algebra) $\mathrm{coend}(F)$ defined in Section 15.2.2, one does not need to start from an *abelian* category, not even from a *linear* one. This is only necessary to ensure that the corresponding evaluation functor (15.2) defines an equivalence, as in Theorem 15.2. For the actual construction of the bialgebra (resp. Hopf algebra) $\mathrm{coend}(F)$, it suffices to start with a monoidal (resp. rigid monoidal category), and these can be defined via generators and relations, much in the same way as monoids and groups.

In many situations, Hopf algebras appear which are defined abstractly via a universal property (for some examples, see Section 15.8), and we hope to convince the reader that it is often a good idea to try and express them in the form $\mathrm{coend}_{\mathcal{C}}(F)$, for a rigid monoidal category $\mathcal{C}$ defined via generators and relations. This allows for much greater flexibility since there are many ways of changing the pair $(\mathcal{C}, F)$ that do not influence the resulting Hopf algebra. In the following sections this philosophy will be applied to the study of the universal bialgebras and Hopf algebras, introduced earlier in this book.

15.3.1 *Algebra Presentations for* $\mathrm{coend}(F)$

We now consider presentations for a (strict) monoidal category $\mathcal{C}$, and use the following notation:

$$\mathcal{C} = \langle (X_k)_k \mid (\phi_l)_l \mid (\chi_m)_m \rangle_{\otimes}.$$

Here, $(X_k)_k$ denote the $\otimes$-generating objects, $(\phi_l)_l$ the $\otimes$-generating morphisms, and $(\chi_m)_m$ the relations among the morphisms. Also, let $F : \mathcal{C} \to \mathrm{vect}_{\mathbb{K}}$ denote a monoidal functor. Then $\mathrm{coend}(F)$ is a bialgebra by Proposition 15.4, and an algebra presentation can be obtained as follows:

(1) If $\mathrm{Ob}(\mathcal{C})$ is not a free monoid on the generating objects $(X_k)_k$, then change the presentation of $\mathcal{C}$ by adding isomorphisms (both arrows and relations) to reduce to this case.

(2) Choose bases $(e_{ki})_i$ for each $F(X_k)$.

(3) The corresponding "matrix coefficients" $(z_{kij})_{kij} \in \mathrm{coend}(F)$ are defined via the coaction

$$\delta(e_{ki}) = \sum_j z_{kij} \otimes e_{kj} \ . \tag{15.8}$$

They generate $\mathrm{coend}(F)$ as an algebra.

(4) Writing out the compatibility of (15.8) with the generating morphisms $(\phi_l)_l$ produces the relations among the generators.

(5) The comultiplication and counit are defined via

$$\Delta(z_{kij}) = \sum_p z_{kip} \otimes z_{kpj}$$
$$\epsilon(z_{kij}) = \delta_{ij}$$

Note that the relations $(\chi_l)_l$ are not used. In practice this process can often be shortened by clever combinatorics. Finally, the procedure outlined above is a more explicit version of formula (15.6) for monoidal categories. In particular, if there are only finitely many X_k and ϕ_l, then one obtains a finite presentation for $\mathrm{coend}(F)$. In the following sections this will be applied to $\underline{\mathrm{end}}(A)$ and $\underline{\mathrm{gl}}(A)$ defined in Chapters 5 and 8.

15.3.2 *Tannakian Reconstruction of* $\underline{\mathbf{end}}(A)$

Assume $A = TV/(R)$, for $R \subset V \otimes V$ and $\dim_{\mathbb{K}}(V) < \infty$, i.e., A is a quadratic algebra. Remember that in Section 6.3, it was shown that the universal bialgebra $\underline{\mathrm{end}}(A)$ coacting on A has the following presentation (where $\bullet$ denotes the black product, introduced in Chapter 4):

$$\underline{\mathrm{end}}(A) = A^! \bullet A = \frac{T(V^* \otimes V)}{\left(\sigma_{(23)}(R^\perp \otimes R)\right)}, \tag{15.9}$$

where

$$R^\perp = \{\phi \in (V \otimes V)^* \mid \phi(R) = 0\},$$

and

$$\sigma_{(23)} \colon V^* \otimes V^* \otimes V \otimes V \to V^* \otimes V \otimes V^* \otimes V$$

transposes the second and third factors. The structure maps Δ and ϵ are given by the usual matrix comultiplication and counit.

Consider the monoidal category

$$\mathcal{C} = \langle r_1, r_2 \mid r_2 \to r_1 \otimes r_1 \rangle_\otimes$$

and the monoidal functor $F \colon \mathcal{C} \to \mathrm{vect}_{\mathbb{K}}$, which is uniquely determined by

$$r_1 \mapsto V,$$
$$r_2 \mapsto R,$$
$$(r_2 \to r_1 \otimes r_1) \mapsto (R \hookrightarrow V \otimes V).$$

15.6 Proposition. *There is an isomorphism of bialgebras*

$$\mathrm{coend}(F) \cong \underline{\mathrm{end}}(A)\ .$$

Proof. This follows from implementing the procedure described in Section 15.3.1 or by using formula (15.6). Let us try the second approach:

$$\mathrm{coend}(F) = \frac{\bigoplus_{X \in \mathcal{C}} FX^* \otimes FX}{E} = \frac{T\big(F(v)^* \otimes F(v)\big)}{E}\ , \tag{15.10}$$

where in the second equality we used that $F(r) \subset F(v) \otimes F(v)$, so we only need to consider sums of powers of $F(v)^* \otimes F(v)$. Now because F is monoidal, the numerator of (15.10) reduces to $T(V^* \otimes V)$. To compute E, it again suffices to consider the generating morphism $r_2 \to r_1 \otimes r_1$, and we see that E is generated by elements

$$y_* \otimes x - y_*|_R \otimes x\ , \quad \text{where } y_* \in (V \otimes V)^* \text{ and } x \in R\ , \tag{15.11}$$

where we identified $x \in R$ with its image under the inclusion $R \hookrightarrow V \otimes V$. This description clearly shows that $E = \big(\sigma_{(23)}(R^\perp \otimes R)\big)$. Finally, one easily checks that the bialgebra structures for $\mathrm{coend}(F)$ and $\underline{\mathrm{end}}(A)$ coincide. □

15.3.3 *Tannakian Reconstruction of* $\underline{\mathrm{gl}}(A)$

At this point we have only provided an alternative construction of $\underline{\mathrm{end}}(A)$, but have not gained anything. This situation changes if one considers $\underline{\mathrm{gl}}(A)$, as defined in Section 8.5. It was shown there that $\underline{\mathrm{gl}}(A)$ can be constructed from $\underline{\mathrm{end}}(A)$ by formally adding an infinite number of generators and relations. It is however not clear that this gives rise to a finitely generated Hopf algebra, or even that $\underline{\mathrm{gl}}(A)$ does not collapse.

In order to ensure that $\underline{\mathrm{gl}}(A)$ has good properties, we restrict our class of quadratic algebras to Koszul Frobenius algebras, which were already considered in Chapter 9. In fact, we will consider their Koszul duals which are more natural from our viewpoint.

15.7 Definition. A connected graded algebra A is *Artin–Schelter* (AS) *regular* of dimension d if it has finite global dimension d, and

$$\mathrm{Ext}^i_A(\mathbb{K}, A) = \begin{cases} 0 & \text{if } i \neq d \\ \mathbb{K}(l) & \text{if } i = d, \end{cases}$$

for some $l \in \mathbb{Z}$ called the *AS-index.*

15.8 *Remark.* This definition requires a few comments:

(1) Many authors also ask for finite GK-dimension (i.e., A is of polynomial growth), but we will not need it in what follows.
(2) For AS-regular algebras the AS-index $l > 0$, see [62, Proposition 3.1].
(3) One can define both left and right AS-regular algebras, but it turns out that the definition is left–right symmetric, see [33, Proposition 2.6]. In particular, Definition 15.7 is unambiguous.

15.9 Lemma ([60, Proposition 5.10]). *Suppose A is Koszul and of finite global dimension, then A is AS-regular if and only if its Koszul dual $A^!$ is Frobenius.*

Proof. The Koszul resolution of A looks like

$$K_\bullet(A)\colon 0 \to A \otimes_{\mathbb{K}} (A^!_d)^* \to \cdots \to A \otimes_k (A^!_1)^* \to A \to \mathbb{K} \to 0\,, \qquad (15.12)$$

which is finite since A has finite global dimension, and in particular $A^! \cong \mathrm{Ext}^*_A(\mathbb{K}, \mathbb{K})$ is finite-dimensional, with top nonzero degree equal to d.

From Definition 15.7, A is AS-regular if and only if the Koszul complex (15.12) is isomorphic to the complex

$$0 \to A \to A^!_1 \otimes A \to \cdots \to A^!_d \otimes A \to 0$$

as complexes of right A-modules. Using the explicit description of the Koszul differentials, this condition is equivalent to the existence of an isomorphism of left A-modules $A^! \to (A^!)^*[-d]$. The existence of such an isomorphism is in turn equivalent to $A^!$ being Frobenius and hence we are done.

In Section 8.5, $\underline{\mathrm{gl}}(A)$ was introduced as the Hopf envelope of $\underline{\mathrm{end}}(A)$, which was denoted $H(\underline{\mathrm{end}}(A))$. We will make use of a categorical analogue of this notion.

15.10 Proposition ([51, Lemma 4.2]). *Let $\mathcal{C}$ be a monoidal category. Then there exists a unique monoidal category $\mathcal{C}^*$ admitting right duals and a monoidal functor $*\colon \mathcal{C} \to \mathcal{C}^*$ such that for any small monoidal category $\mathfrak{D}$ admitting right duals and monoidal functor F, there exists a unique monoidal functor F^* making the diagram*

$$\begin{array}{ccc} \mathcal{C} & \xrightarrow{*} & \mathcal{C}^* \\ {\scriptstyle F}\downarrow & \swarrow_{F^*} & \\ \mathcal{D} & & \end{array}$$

commute. Moreover, this construction is compatible with Hopf envelopes: for any monoidal category $\mathcal{C}$ *and monoidal functor* $F\colon \mathcal{C} \to \mathrm{vect}_{\mathbb{K}}$, *there are isomorphisms of Hopf algebras*

$$H\big(\mathrm{coend}_{\mathcal{C}}(F)\big) \cong \mathrm{coend}_{\mathcal{C}^*}(F^*)\,.$$

15.11 Corollary. *For* $(\mathcal{C}, F)$ *as in Section* 15.3.2, *there is an isomorphism*

$$\underline{\mathrm{gl}}(A) \cong \mathrm{coend}_{\mathcal{C}^*}(F^*)$$

of Hopf algebras.

Hence, to obtain a presentation for $\underline{\mathrm{gl}}(A)$, it suffices to construct *rigidisations* of $(\mathcal{C}, F)$. Because one is only interested in the resulting coend, there is again a lot of extra freedom. The following proposition illustrates the end result of such a construction, and provides a minimal presentation of $\underline{\mathrm{gl}}(A)$.

15.12 Proposition ([54, App. A]). *For* $A = TV/(R)$ *a Koszul, Artin–Schelter regular algebra of global dimension* d, *consider the monoidal category*

$$\mathcal{D} = \langle r_1, r_2, \ldots, r_{d-1}, r_d^{\pm 1} \mid r_i \to r_1^{\otimes i},\ r_a r_d^{-1} r_{d-a} \to 1\rangle_{\otimes}\,,$$

where i *runs over* $\{2, \ldots, d\}$ *and* $a \in \{1, \ldots, d-1\}$ *is fixed (so there are* d *generating morphisms). Define a monoidal functor* $G\colon \mathcal{D} \to \mathrm{vect}_k$ *via*

$$\begin{aligned} G(r_1) &= V \\ G(r_i) &= \bigcap_{k+l+2=i} V^{\otimes k} \otimes R \otimes V^{\otimes l} \quad \textit{for } i \geq 2 \\ G(r_i \to r_1^{\otimes i}) &= \left(\bigcap_{k+l+2=i} V^{\otimes k} \otimes R \otimes V^{\otimes l} \hookrightarrow V^{\otimes i} \right) \end{aligned}$$

Then $\mathrm{coend}_{\mathcal{D}}(G) \cong \underline{\mathrm{gl}}(A)$.

Note that $\mathcal{D}$ only has a finite number of generating objects, so in particular, $\underline{\mathrm{gl}}(A)$ is finitely generated. This is a consequence of AS-regularity: it ensures that $\dim(A^!_d) = 1$, and that the obvious inclusions

$$(A^!_d)^* \hookrightarrow (A^!_a)^* \otimes (A^!_{d-a})^*$$

define nondegenerate pairings between $(A^!_a)^*$ and $(A^!_{d-a})^*$ (cf. Lemma 15.9). This ensures one only has to formally invert a single object in order to construct a pair $(\mathcal{D}, G)$ such that

$$\mathrm{coend}_{\mathcal{D}}(G) \cong \mathrm{coend}_{\mathcal{C}^*}(F^*) \cong \underline{\mathrm{gl}}(A)\,. \tag{15.13}$$

Using Proposition 15.12 one can, in fact, check that $\underline{\mathrm{gl}}(A)$ is generated by $(z_{ij})_{i,j=1}^{d}$ (corresponding to the matrix coefficients of r_1) and the inverse of a group-like element δ (corresponding to the matrix coefficient of r_d). The relations among these generators are somewhat cumbersome to write down explicitly so we illustrate them in a special case.

15.13 *Example.* Let $A = \mathbb{K}[x, y]$, then

$$R \subset V \otimes V : r \mapsto x \otimes y - y \otimes x\,,$$

and $d = 2$. Applying the procedure outlined in Section 15.3.1 to $(\mathcal{D}, G)$ from Proposition 15.12 we find that $\underline{\mathrm{aut}}(A)$ is generated as an algebra, by a, b, c, d, δ^{-1} with the following relations

$$\begin{aligned} ac - ca = 0 &= bd - db\,,\\ ad - cb = \delta &= da - bc\,,\\ \delta\delta^{-1} = 1 &= \delta^{-1}\delta\,,\\ a\delta^{-1}d - b\delta^{-1}c = 1 &= d\delta^{-1}a - c\delta^{-1}b\,,\\ b\delta^{-1}a - a\delta^{-1}b = 0 &= c\delta^{-1}d - d\delta^{-1}c\,. \end{aligned} \tag{15.14}$$

Unsurprisingly $\underline{\mathrm{gl}}(A)$, like $\underline{\mathrm{end}}(A)$, has exponential growth.

15.4 Highest Weight Categories and Quasi-Hereditary Coalgebras

The Hopf algebra $\underline{\mathrm{gl}}(A)$ satisfies the following universal property, analogous to the one for $\underline{\mathrm{end}}(A)$ considered in Chapters 5 and 6.

15.14 Proposition. *If H is a Hopf algebra and A is an H-comodule algebra given by $f : A \to H \otimes A$ such that $f(A_n) \subset H \otimes A_n$, then there is a unique morphism of Hopf algebras $g : \underline{\mathrm{gl}}(A) \to H$ such that the diagram*

$$\begin{array}{ccc} A & \xrightarrow{\delta_A} & \underline{\mathrm{gl}}(A) \otimes A \\ & \searrow^{f} & \downarrow g \otimes 1 \\ & & H \otimes A \end{array}$$

commutes.

For $A = \mathrm{Sym}_{\mathbb{K}}(V)$ a polynomial ring, Proposition 15.14 ensures that the coaction

$$A \to \mathcal{O}\big(\mathrm{GL}(V)\big) \otimes A$$

induced by the standard action of $\mathrm{GL}(V)$ on V, factors through $\underline{\mathrm{gl}}(A)$. Hence, there is a natural functor

$$\mathrm{comod}\left(\mathcal{O}\big(\mathrm{GL}(V)\big)\right) \to \mathrm{comod}\big(\underline{\mathrm{gl}}(A)\big) ,$$

and since (as we saw in Example 15.5) $\mathrm{comod}\left(\mathcal{O}\big(\mathrm{GL}(V)\big)\right)$ is equivalent to $\mathrm{rep}_{\mathbb{K}}\big(\mathrm{GL}(V)\big)$, this suggests that $\underline{\mathrm{gl}}(A)$ has an interesting representation theory.

In arbitrary characteristic, $\mathrm{rep}_{\mathbb{K}}\big(\mathrm{GL}(V)\big)$ is an important example of a highest weight category (see Section 15.4.2 for a definition), and the main result of [54] asserts that $\mathrm{comod}\big(\underline{\mathrm{gl}}(A)\big)$ is also a highest weight category, for any Koszul, Artin–Schelter regular algebra A.

15.4.1 *Representations of the General Linear Group*

Consider the category $\mathrm{rep}_{\mathbb{K}}(G)$ of rational (right) representations[3] of $G = \mathrm{GL}(V)$, for $\dim(V) = n$. By definition, a representation X of G is rational if for some (and hence every) basis $e_1, \ldots, e_m$ of X,

$$e_i \cdot g = \sum_{j=1}^{m} e_j f_{ij}(g) , \quad \text{for all } g \in G \text{ and } i = 1, \ldots, m, \tag{15.15}$$

for some coefficient functions $f_{ij} \in \mathcal{O}(G)$. A rational representation X can be given the structure of a left comodule by defining the coaction to be

$$\delta \colon X \to \mathcal{O}(G) \otimes X : e_i \mapsto \sum_{j=1}^{m} f_{ij} \otimes e_j ,$$

[3] As before, any representation is by definition finite-dimensional.

and this association defines an equivalence of categories between finite-dimensional $\mathcal{O}(G)$-comodules and rational representations of G.

The inclusion of monoids $G \hookrightarrow M_n(\mathbb{K})$ induces an inclusion of bialgebras

$$\mathcal{O}(M_n) \hookrightarrow \mathcal{O}(G)\colon \mathbb{K}[x_{11}, \ldots, x_{nn}] \hookrightarrow \mathbb{K}[x_{11}, \ldots, x_{nn}, \det^{-1}],$$

where $x_{ij}(g)$ is the ijth entry of the matrix g and det is the determinant function. Then a rational representation $X \in \operatorname{rep}_{\mathbb{K}}(G)$ is called *polynomial* if the coefficient functions f_{ij} in (15.15) all belong to $\mathcal{O}(M_n)$. Now for any representation $V \in \operatorname{rep}_{\mathbb{K}}(G)$, $V \otimes \left(\bigwedge^n(V)\right)^{\otimes m}$ is polynomial for some $m \geq 0$, so we can restrict to studying polynomial representations of G, or equivalently, $\mathcal{O}(M_n)$-comodules.

If $\operatorname{char}(\mathbb{K}) = 0$, it is classical that $\operatorname{rep}_{\mathbb{K}}(G)$ is semisimple: every representation in $\operatorname{rep}_{\mathbb{K}}(G)$ is a direct sum of simple representations. The simple polynomial representations $L(\lambda)$ are classified by the set of partitions with at most n rows

$$\Lambda = \{\lambda = (\lambda_1, \ldots, \lambda_n) \in \mathbb{N}^n \,|\, \lambda_1 \geq \lambda_2 \geq \cdots \geq \lambda_n\}.$$

This collection can be upgraded to a poset by setting

$$\lambda \leq \mu \iff \sum_{j=1}^{k} \lambda_j \leq \sum_{j=1}^{k} \mu_j \text{ for all } k = 1, \ldots, n.$$

Now consider the subgroups $T \subset B \subset G$ of diagonal and lower triangular matrices. Denoting the simple representation corresponding to $\lambda \in \Lambda$ by $L(\lambda)$, these simple representations can be explicitly constructed as

$$L(\lambda) = \operatorname{ind}_B^G(\lambda),$$

i.e., one considers λ as a one-dimensional representation of T, extends it to a B-representation by letting the unipotent part act trivially, and then induces to a representation of G.

15.15 *Example.* If $\dim(V) = 2$, then for $\lambda = (\lambda_1, \lambda_2) \in \Lambda$, we have

$$L(\lambda) = \operatorname{ind}_B^G(\lambda) = \operatorname{Sym}^{\lambda_1 - \lambda_2}(V) \otimes \textstyle\bigwedge^2(V)^{\otimes \lambda_2},$$

so any simple rational G-representation is isomorphic to the tensor product of an integer power of the determinant representation and a symmetric power of the tautological representation.

In positive characteristic the situation is not nearly as easy, as the following simple example shows.

15.16 *Example.* Assume that $\mathrm{char}(k) = p > 0$, and $V = ke_1 + ke_2$. Then

$$\mathrm{ind}_B^G((p,0)) = \mathrm{Sym}^p(V) = \sum_{i=0}^{p} k e_1^{p-i} e_2^i$$

contains the two-dimensional simple subrepresentation $L = \mathbb{K}e_1^p + \mathbb{K}e_2^p$. Note that L is not even a direct summand of $\mathrm{Sym}^p(V)$.

In particular, we see that $\mathrm{rep}_{\mathbb{K}}(G)$ is no longer semisimple. It is, however, still true that the simple representations are classified by Λ, and occur as subrepresentations of the induced representations. For any $\lambda \in \Lambda$, denote $\lambda^* := -w_0\lambda$, where w_0 denotes the longest element in the Weyl group.

15.17 Theorem. *For any simple polynomial representation L of G, there is a unique $\lambda \in \Lambda$ such that*

$$\begin{aligned} L &\cong \mathrm{soc}\big(\mathrm{ind}_B^G(\lambda)\big) \\ &\cong \mathrm{top}\big(\mathrm{ind}_B^G(\lambda^*)^*\big)\,. \end{aligned}$$

The simple representation corresponding to λ is denoted $L(\lambda)$.

This seems like good news, but in fact this information is not even explicit enough to determine the characters of the simple representations. Indeed, to determine these characters is one of the main motivating problems in the field of modular representation theory of reductive algebraic groups.

The theorem does, however, suggest that the induced representations $\mathrm{ind}_B^G(\lambda)$ and $\mathrm{ind}_B^G(\lambda^*)^*$ still play an important role. For this reason they are denoted $\nabla(\lambda)$ (respectively $\Delta(\lambda)$) and are called *costandard* (respectively *standard*) representations. We will mostly focus on the costandard representations, since the standard ones are their dual (for a precise statement, see Proposition 15.33). Their characters can be computed using the Weyl character formula, and they satisfy the following important properties.

15.18 Proposition. *For all $\lambda \in \Lambda$:*

(1) $\mathrm{End}_G\big(\nabla(\lambda)\big) \cong \mathbb{K}$,
(2) $\mathrm{Hom}_G\big(\nabla(\lambda), \nabla(\mu)\big) \neq 0 \Rightarrow \lambda \geq \mu$,
(3) $\mathrm{Ext}_G^1\big(\nabla(\lambda), \nabla(\mu)\big) \neq 0 \Rightarrow \lambda > \mu$.
(4) $\mathrm{Ext}_G^i\big(\Delta(\lambda), \nabla(\mu)\big) = 0$ *for* $i > 0$.

15.19 Corollary. *Denoting by* $[- : -]$ *composition multiplicities, we find that* $[\nabla(\lambda) : L(\lambda)] = 1$ *and if* $[\nabla(\lambda) : L(\mu)] \neq 0$, *then* $\lambda \geq \mu$.

This, in turn, indicates that the category of representations filtered by the $\nabla(\lambda)$ plays an important role. Denote by $\mathcal{F}(\nabla)$ the exact subcategory of $\mathrm{rep}_{\mathbb{K}}(G)$ consisting of representations filtered by the $\nabla(\lambda)$, for $\lambda \in \Lambda$. We similarly define $\mathcal{F}(\Delta)$. The following deep result is due to Mathieu.

15.20 Theorem ([48]). *If $X, Y \in \mathcal{F}(\nabla)$, then also $X \otimes_K Y \in \mathcal{F}(\nabla)$. The same result holds for $\mathcal{F}(\Delta)$.*

A last class of modules which we will need, and which play an important role in the representation theory of $\mathrm{GL}(V)$ are the *tilting modules.*

15.21 Definition. The category $\mathcal{F}(\nabla) \cap \mathcal{F}(\Delta)$ of all modules having both a ∇-filtration and a Δ-filtration is called the category of *tilting modules.*

The name is explained by the following proposition, which follows immediately from Proposition 15.18(4).

15.22 Proposition. *For any $M \in \mathcal{F}(\nabla) \cap \mathcal{F}(\Delta)$,*

$$\mathrm{Ext}^i_G(M, M) = 0\,, \quad \text{for all } i > 0.$$

15.23 Proposition. *For any indecomposable representation $T \in \mathcal{F}(\nabla) \cap \mathcal{F}(\Delta)$, there is a unique $\lambda \in \Lambda$ and exact sequences*

$$0 \to K(\lambda) \to T \to \nabla(\lambda) \to 0\,,$$

and

$$0 \to \Delta(\lambda) \to T \to K'(\lambda) \to 0$$

such that $K(\lambda)$ (respectively $K'(\lambda)$) has a filtration by $\nabla(\mu)$ (respectively $\Delta(\mu)$) with $\mu < \lambda$. This T is denoted $T(\lambda)$.

In particular, we find that

$$\mathrm{add}\left(\bigoplus_{\lambda \in \Lambda} T(\lambda)\right) = \mathcal{F}(\nabla) \cap \mathcal{F}(\Delta)\,,$$

where $\mathrm{add}(M)$ denotes the category consisting of all representations isomorphic to direct summands of direct sums of M, and this module $T = \bigoplus_{\lambda \in \Lambda} T(\lambda)$ is called the *characteristic tilting module* (which is infinite-dimensional).

15.24 *Example.* For $\dim(V) = 2$ and $i < p$, one has

$$L\big((i,0)\big) = \nabla\big((i,0)\big) = \Delta\big((i,0)\big) = T\big((i,0)\big) = S^i(V)\,.$$

In general, these tilting representations are also hard to describe, but they can be related to more familiar representations. In fact,

$$\bigwedge^i V = \nabla\big(\lambda(i)\big) = \Delta\big(\lambda(i)\big) = L\big(\lambda(i)\big) = T\big(\lambda(i)\big)$$

for $i = 0, \ldots, n$ and $\lambda(i) = (1, \ldots, 1, 0, \ldots, 0) \in \Lambda$ with 1 appearing i times. Hence, by Theorem 15.20 all tensor products of the $\bigwedge^i(V)$ are tilting representations.

15.25 Theorem. *For $\lambda = (\lambda_1, \ldots, \lambda_n) \in \Lambda$, there is a decomposition*

$$\bigwedge^{\lambda_1^t} V \otimes_{\mathbb{K}} \cdots \otimes_k \bigwedge^{\lambda_l^t} V = T(\lambda) \oplus Y\ ,$$

where Y is a direct sum of tilting representations $T(\mu)$ with $\mu < \lambda$, and $\lambda^t = (\lambda_1^t, \ldots, \lambda_l^t)$ is the conjugate partition.

If we denote by

$$\mathcal{V} = \big\langle \bigwedge^i V \mid i = 1, \ldots, n \big\rangle_{\otimes} \subset \mathrm{rep}_{\mathbb{K}}(G)$$

the full monoidal subcategory generated by the exterior powers of the standard representation, and by $F\colon \mathcal{V} \to \mathrm{vect}_{\mathbb{K}}$ the restriction of the forgetful functor, then

$$\mathrm{coend}(F) \cong \mathcal{O}(G)\ .$$

Let $\mathrm{Perf}(\mathcal{V}^{\mathrm{op}})$ be the triangulated category of finite complexes of finitely generated projective right $\mathcal{V}$-modules. Then $\mathrm{Perf}(\mathcal{V}^{\mathrm{op}})$ has a natural structure of monoidal triangulated category by putting

$$\mathcal{V}(-, u) \otimes \mathcal{V}(-, v) = \mathcal{V}(-, uv)$$

and extending to complexes. The functor F extends to an exact monoidal functor

$$F\colon \mathrm{Perf}(\mathcal{V}^{\mathrm{op}}) \to \mathbf{D}^b\big(\mathrm{rep}_{\mathbb{K}}(G)\big)\colon \mathcal{V}(-, u) \mapsto F(u)\ . \tag{15.16}$$

15.26 *Remark.* At the risk of confusing various tensor products the functor M can be written intrinsically as $- \overset{L}{\otimes}_{\mathcal{V}} M$.

15.27 Theorem. *The functor*

$$F\colon \mathrm{Perf}(\mathcal{V}^{\mathrm{op}}) \to \mathbf{D}^b\big(\mathrm{rep}_k(G)\big)\colon \mathcal{V}(-, u) \mapsto F(u)\ .$$

is an equivalence of monoidal triangulated categories.

Proof. This follows from combining Theorem 15.25, Proposition 15.23, and Corollary 15.19.

Note that the tensor generators for $\mathcal{V}$, which govern the derived category of $\operatorname{rep}_{\mathbb{K}}(G)$ by Theorem 15.27, correspond to the terms in the Koszul resolution

$$0 \to A \otimes \bigwedge^n V \to \cdots \to A \otimes \bigwedge^l V \to \cdots \to A \otimes V \to A \to \mathbb{K} \to 0$$

of $A = \operatorname{Sym}_{\mathbb{K}}(V)$. We will use this in Section 15.5 as a starting point to study the representation theory of $\underline{\mathrm{gl}}(A)$, for an arbitrary Koszul, Artin–Schelter regular algebra A.

15.4.2 *Highest Weight Categories and Quasi-Hereditary Coalgebras*

A lot of the structure present in $\operatorname{rep}_{\mathbb{K}}(G)$ can be formalized and gives rise to the notion of a *highest weight category*.[4] It is this notion that we will be able to carry over to the noncommutative setting and $\underline{\mathrm{gl}}(A)$. Remember that a poset $(\Lambda, \leq)$ is called *interval finite* if for any $\lambda, \mu \in \Lambda$, the set $\{\psi \in \Lambda \mid \lambda \leq \psi \leq \mu\}$ is finite.

To emphasize the analogy with algebraic groups, whenever dealing with coalgebras, we will use the term *representation* as a synonym for a finite-dimensional comodule.

15.28 Definition. Let C be a coalgebra and let $(\Lambda, \leq)$ be an interval finite poset indexing the simple representations. Then C is *quasi-hereditary* if there exist finite-dimensional comodules $\nabla(\lambda)$, for all $\lambda \in \Lambda$, such that:

(1) $\operatorname{End}_C(\nabla(\lambda)) \cong \mathbb{K}$,
(2) $\operatorname{Hom}_C(\nabla(\lambda), \nabla(\mu)) \neq 0 \Rightarrow \lambda \geq \mu$,
(3) $\operatorname{Ext}^1_C(\nabla(\lambda), \nabla(\mu)) \neq 0 \Rightarrow \lambda > \mu$,
(4) C has a filtration with subquotients of the form $\nabla(\lambda)$.

The category $\operatorname{comod}(C)$ is a highest weight category and the comodules $\nabla(\lambda)$ are called *costandard comodules*.

15.29 *Remark.* (1) The poset $(\Lambda, \leq)$ is a part of the data defining a highest weight category. In particular there can be different quasi-hereditary structures on the same underlying coalgebra.

(2) There is no mention yet of a tensor product. Indeed, for an arbitrary coalgebra, $\operatorname{comod}(C)$ is not necessarily monoidal.

---

[4] This notion plays a very important role in Lie theory, see [8, 34]

(3) Definition 15.28 is not quite standard and relies on the standardisation result of Dlab and Ringel [18]. See [55, Appendix A] for a detailed comparison.

15.30 *Example.* If C is a directed coalgebra, i.e., there is an ordering of the simple representations by a poset Λ such that $\mathrm{Ext}^1_C\big(L(\lambda), L(\mu)\big) \neq 0$ implies that $\lambda > \mu$, then one can take $\nabla(\lambda) = L(\lambda)$.

In fact, in some sense a quasi-hereditary coalgebra is a direct generalization of Example 15.30. Indeed, Definition 15.28 implies that the $\nabla(\lambda)$ form a full exceptional collection in $\mathbf{D}^b\big(\mathrm{comod}(C)\big)$.

15.31 Definition. An object X in a $\mathbb{K}$-linear triangulated category $\mathcal{T}$ is called *exceptional* if

$$\mathrm{Hom}_{\mathcal{T}}(X, X[r]) = \begin{cases} \mathbb{K} & \text{if } r = 0, \\ 0 & \text{otherwise.} \end{cases} \tag{15.17}$$

A collection of objects $\{X_i\}_{i\in(I,\leq)}$ in a $\mathbb{K}$-linear triangulated category $\mathcal{T}$, for some poset $(I, \leq)$, is called an *exceptional collection* if each X_i is exceptional and:

$$\mathrm{Hom}_{\mathcal{T}}(X_p, X_q[r]) = 0, \tag{15.18}$$

if $p > q$ and $r \in \mathbb{Z}$. An exceptional collection is called *full* if it generates all of $\mathcal{T}$.

The full exceptional collection $\big(\nabla(\lambda)\big)_{\lambda\in\Lambda}$ is special since it consists solely of objects living in the heart of the standard t-structure.

15.32 Definition. For a full exceptional collection $(F_\lambda)_{\lambda\in(\Lambda,\leq)}$ in a triangulated category $\mathcal{T}$, there is a dual full exceptional collection $(E_\lambda)_{\lambda\in(\Lambda,\leq^{\mathrm{op}})}$, uniquely determined by

$$\mathrm{Hom}(E_\lambda, F_\mu[i]) = \begin{cases} \mathbb{K} & \text{if } \lambda = \mu \text{ and } i = 0, \\ 0 & \text{otherwise.} \end{cases}$$

15.33 Proposition. *For a highest weight category* comod(C), *the exceptional collection dual to the costandard comodules consists again of indecomposable comodules, which are called* standard comodules, *and are denoted* $\Delta(\lambda)$.

There is another definition of a quasi-hereditary coalgebra, which is often easier to work with in practice. We will first need the definition of a heredity chain which we will phrase in the context of finite-dimensional algebras. So assume A is a finite-dimensional $\mathbb{K}$-algebra, with Jacobson radical rad(A).[5]

15.34 Definition. A two-sided ideal I of A is called a *heredity ideal* if

[5] Recall that the *Jacobson radical* of A is the intersection of all annihilators of simple right A-modules.

(1) I is idempotent,
(2) I_A is projective,
(3) $I \operatorname{rad}(A) I = 0$.

15.35 Definition. The algebra A is a quasi-hereditary algebra if it has a filtration by heredity ideals, i.e., there is a chain

$$0 = J_0 \subset J_1 \subset \cdots \subset J_{m-1} \subset J_m = A$$

of ideals of A such that J_t/J_{t-1} is a heredity ideal in A/J_{t-1} for any $1 \leq t \leq m$. Such a chain is called a heredity chain.

15.36 Definition. A (possibly infinite-dimensional) coalgebra C is *quasi-hereditary* if there exists an exhaustive filtration

$$0 \subset C_1 \subset C_2 \subset \cdots \subset C_n \subset \cdots$$

of finite-dimensional subcoalgebras such that for every i, we have

$$0 = (C_i/C_i)^* \subset (C_i/C_{i-1})^* \subset (C_i/C_{i-2})^* \subset \cdots \subset C_i^*$$

is a heredity chain. Such a filtration is called a *heredity cochain*.

15.5 Representations of $\underline{\mathrm{gl}}(A)$

Consider a Koszul, Artin–Schelter regular algebra $A = TV/(R)$ of global dimension d, and the corresponding universal Hopf algebra $\underline{\mathrm{gl}}(A)$. Taking our cue from Theorem 15.27 and the preceding constructions, consider the Koszul resolution of A:

$$0 \to A \otimes R_d \to \cdots \to A \otimes R_l \to \cdots \to A \otimes R \to A \otimes V \to A \to \mathbb{K} \to 0\,,$$

with[6] $R_l := \bigcap_{i+j+2=l} V^i R V^j$. In particular, we have $R_2{=}R$ and for uniformity we also put $R_1 = V$. It follows from the basic properties of AS-regular algebras that $\dim R_d = 1$ and that, moreover, the obvious inclusions $R_d \hookrightarrow R_a R_{d-a}$ define nondegenerate pairings between R_a and R_{d-a}. These properties characterize the AS-regular algebras among the Koszul ones, as we saw in Lemma 15.9.

Since we would like to think of $\underline{\mathrm{gl}}(A)$ as a noncommutative version of the coordinate ring of $\mathrm{GL}(V)$, we will denote $\operatorname{comod}\big(\underline{\mathrm{gl}}(A)\big)$ by $\operatorname{rep}_{\mathbb{K}}\big(\underline{\mathrm{gl}}(A)\big)$. It is easy to

[6] We usually omit tensor product signs.

see that the $(R_l)_l$ are $\underline{\mathrm{gl}}(A)$-comodules, with R_d being invertible. The discussion after Theorem 15.27 suggests to consider

$$\mathcal{V} = \langle R_l \mid l = 1, \ldots, d \rangle_{\otimes} \subset \mathrm{rep}_{\mathbb{K}}\big(\underline{\mathrm{gl}}(A)\big) .$$

Moreover, denoting $F \colon \mathcal{V} \to \mathrm{vect}_{\mathbb{K}}$ the restriction of the forgetful functor, one might *expect* in analogy with the commutative setting that

$$\mathrm{coend}(F) \cong \underline{\mathrm{gl}}(A),$$

and that F induces an equivalence

$$F \colon \mathrm{Perf}(\mathcal{V}^{\mathrm{op}}) \to \mathbf{D}^b\Big(\mathrm{rep}_{\mathbb{K}}\big(\underline{\mathrm{gl}}(A)\big)\Big) : \mathcal{V}(-, v) \mapsto F(v) \tag{15.19}$$

of monoidal triangulated categories. However it seems difficult to verify this directly since the structure of $\mathrm{rep}_{\mathbb{K}}\big(\underline{\mathrm{gl}}(A)\big)$ is completely unknown at this stage. Therefore we proceed differently.

We will relate $\mathcal{V}$ to a certain monoidal category $\mathcal{U}$ with strong combinatorial features. In fact, we have already introduced a suitable monoidal category $\mathcal{D}$ in Proposition 15.12 but the latter was optimized for finding a compact presentation of $\underline{\mathrm{gl}}(A)$. In contrast, $\mathcal{U}$ will more faithfully reflect the representation theoretic features of $\underline{\mathrm{gl}}(A)$.

To summarize: we will not use $(\mathcal{D}, G)$ like in Proposition 15.12 but use a different pair $(\mathcal{U}, M)$. Nonetheless we will have

$$\mathrm{coend}_{\mathcal{U}}(M) \cong \mathrm{coend}_{\mathcal{D}}(G) \cong \underline{\mathrm{gl}}(A).$$

This illustrates the fact that different pairs $(\mathcal{C}, F)$ with the same coend can be used to study different aspects of the same Hopf algebra.

15.5.1 *The Category* $\mathcal{U}$

It is not hard to see that there are morphisms of $\underline{\mathrm{gl}}(A)$-representations

$$\begin{aligned} \Phi_{a,b} &\colon R_{a+b} \to R_a R_b \, , \\ \Theta_{a,b} &\colon R_a R_d^{-1} R_b \to R_{a+b-d} \, , \end{aligned}$$

satisfying certain natural relations. To formalize this, define the monoid

$$\Lambda = \langle r_1, \ldots, r_{d-1}, r_d^{\pm 1} \rangle$$

and consider the following monoidal categories with set of objects Λ:

$$\mathcal{U}_\uparrow = \langle r_1, \ldots, r_{d-1}, r_d^{\pm 1} \mid \phi_{a,b} \colon r_{a+b} \to r_a r_b \rangle_\otimes \,,$$

and impose the following set of relations:

$$\begin{array}{ccc} r_{a+b+c} & \xrightarrow{\phi_{a,b+c}} & r_a r_{b+c} \\ \Big\downarrow{\scriptstyle \phi_{a+b,c}} & & \Big\downarrow{\scriptstyle r_a \phi_{b,c}} \\ r_{a+b} r_c & \xrightarrow{\phi_{a,b} r_c} & r_a r_b r_c \end{array} \tag{15.20}$$

writing u for Id_u and suppressing tensor products as usual. Similarly, consider

$$\mathcal{U}_\downarrow = \langle r_1, \ldots, r_{d-1}, r_d^{\pm 1} \mid \theta_{a,b} \colon r_a r_d^{-1} r_b \to r_{a+b-d} \rangle_\otimes$$

and impose the relations:

$$\begin{array}{ccc} r_a r_d^{-1} r_b r_d^{-1} r_c & \xrightarrow{\theta_{a,b} r_d^{-1} r_c} & r_{a+b-d} r_d^{-1} r_c \\ \Big\downarrow{\scriptstyle r_a r_d^{-1} \theta_{b,c}} & & \Big\downarrow{\scriptstyle \theta_{a+b-d,c}} \\ r_a r_d^{-1} r_{b+c-d} & \xrightarrow{\theta_{a,b+c-d}} & r_{a+b+c-2d} \end{array} \tag{15.21}$$

Relations (15.20) and (15.21) are chosen because they are satisfied by the morphisms $\Phi_{a,b}$ and $\Theta_{a,b}$.

Now let $\widetilde{\mathcal{U}} = \mathcal{U}_\downarrow * \mathcal{U}_\uparrow$ be the category with set of objects Λ and the morphisms freely generated by the morphisms in $\mathcal{U}_\downarrow$ and $\mathcal{U}_\uparrow$. Then $\widetilde{\mathcal{U}}$ is strict monoidal in the obvious way. Let $\mathcal{U}$ be the monoidal quotient of $\widetilde{\mathcal{U}}$ obtained by imposing the following sets of relations

(1)

$$\begin{array}{ccc} r_{a+b}r_d^{-1}r_c & \xrightarrow{\phi_{a,b}r_d^{-1}r_c} & r_ar_br_d^{-1}r_c \\ \downarrow \theta_{a+b,c} & & \downarrow r_a\theta_{b,c} \\ r_{a+b+c-d} & \xrightarrow{\phi_{a,b+c-d}} & r_ar_{b+c-d} \end{array} \tag{15.22}$$

where $d \leq b + c$ and where moreover we allow the degenerate cases $a + b = d$ in which case we put $\theta_{d,c} = \mathrm{Id}_{r_c}$ and $b + c = d$ in which case we put $\phi_{a,0} = \mathrm{Id}_{r_a}$.

(2)

$$\begin{array}{ccc} r_ar_d^{-1}r_{b+c} & \xrightarrow{r_ar_d^{-1}\phi_{b,c}} & r_ar_d^{-1}r_br_c \\ \downarrow \theta_{a,b+c} & & \downarrow \theta_{a,b}r_c \\ r_{a+b+c-d} & \xrightarrow{\phi_{a+b-d,c}} & r_{a+b-d}r_c \end{array} \tag{15.23}$$

where $d \leq a + b$ and where, moreover, we allow the degenerate cases $b + c = d$ when we put $\theta_{a,d} = \mathrm{Id}_{r_a}$, and $a + b = d$ when we put $\phi_{0,c} = \mathrm{Id}_{r_c}$.

Again, it is easy to see that the $\Phi_{a,b}$ and $\Theta_{a,b}$ satisfy relations (15.22) and (15.23).

By linearizing the morphism spaces in these monoidal categories, we obtain linear categories $\mathbb{K}\mathcal{U}_\uparrow, \mathbb{K}\mathcal{U}_\downarrow$, and $\mathbb{K}\mathcal{U}$. It is possible to put a grading on the morphisms of these categories, so one can consider them as multiple object versions of graded algebras. The combinatorial structure of $\mathcal{U}$ is elucidated in the following proposition.

15.37 Proposition ([54, Propositions 3.1.2, 3.3.1]).

(1) *The graded categories* $\mathbb{K}\mathcal{U}_\uparrow$ *and* $\mathbb{K}\mathcal{U}_\downarrow$ *are Koszul,*

(2) $\mathcal{U}$ *can be given the structure of a* Reedy category, *i.e., every morphism* f *in* $\mathcal{U}$ *can be written uniquely as a composition* $f_\uparrow \circ f_\downarrow$ *with* $f_\downarrow$ *in* $\mathcal{U}_\downarrow$ *and* $f_\uparrow$ *in* $\mathcal{U}_\uparrow$.

15.5.2 $\underline{\mathrm{gl}}(A)$ *is Quasi-Hereditary*

Since relations (15.20)–(15.23) were chosen based on the relations satisfies by the $\Phi_{a,b}$ and $\Theta_{a,b}$, it follows that by construction there is a monoidal functor $G\colon \mathcal{U} \to \mathcal{V}$, which can be composed with the forgetful functor F to obtain a monoidal functor

$$M\colon \mathcal{U} \to \mathrm{vect}_{\mathbb{K}}\,. \tag{15.24}$$

Note that up to some morphisms and relations, this setup is very similar to Proposition 15.12, so the following theorem should not come as a surprise.

15.38 Theorem ([54, Theorem 5.1]). *The monoidal category $\mathcal{U}$ is rigid, and there is an isomorphism of Hopf algebras*

$$\mathrm{coend}_{\mathcal{U}}(M) \cong \underline{\mathrm{gl}}(A).$$

At this point we forget about the intermediate category $\mathcal{V}$, which is a priori hard to control since it is a linear category, and focus only on $\mathcal{U}$, which is not linear, and on the functor M. This turns out to greatly simplify calculations, and is loosely analogous to using group theory instead of ring theory to study a group algebra.

Equipping Λ with the left- and right-invariant partial ordering generated by

$$r_{a+b} < r_a r_b,$$
$$r_{d-a-b} < r_{d-a} r_d^{-1} r_{d-b},$$

we can now state the main theorem from [54] more precisely.

15.39 Theorem. *The coalgebra $\underline{\mathrm{gl}}(A)$ is quasi-hereditary with respect to the poset $(\Lambda, \leq)$. The costandard and standard representations are given as*

$$\nabla(\lambda) = \mathrm{coker}\left(\bigoplus_{\substack{\mu \to \lambda \text{ in } \mathcal{U} \\ \mu < \lambda}} M(\mu) \to M(\lambda)\right)$$
$$\Delta(\lambda) = \ker\left(M(\lambda) \to \bigoplus_{\substack{\lambda \to \mu \text{ in } \mathcal{U} \\ \mu < \lambda}} M(\mu)\right). \tag{15.25}$$

The proof of this theorem uses the strong combinatorial structure on the category $\mathcal{U}$ from Proposition 15.37 in order to check Definition 15.36. Assuming that $\Lambda_1 \subset \Lambda_2$ are saturated subsets[7] of Λ such that the elements of $\Lambda_2 - \Lambda_1$ are incomparable. Let $\mathcal{U}_i \subset \mathcal{U}$ be the full subcategories of $\mathcal{U}$ with object sets Λ_i. The key technical result that enters in the proof of Theorem 15.39 is the following.

15.40 Theorem. *There is an exact sequence*

$$0 \to \prod_{\lambda \in \Lambda_2 - \Lambda_1} \mathrm{Hom}_{\mathbb{K}}\big(\nabla(\lambda), \Delta(\lambda)\big) \to \mathrm{End}_{\mathcal{U}_2}(M) \to \mathrm{End}_{\mathcal{U}_1}(M) \to 0\,, \tag{15.26}$$

where $\nabla(\lambda)$ and $\Delta(\lambda)$ are as in (15.25) *above.*

Starting with (15.26) we may construct a heredity cochain in $\mathrm{coend}_{\mathcal{U}}(M)$ which yields that $\mathrm{coend}_{\mathcal{U}}(M) \cong \underline{\mathrm{gl}}(A)$ is quasi-hereditary. The following theorem provides an analogue of Theorem 15.20.

[7] A subset $\pi \subset \Lambda$ is called *saturated* if $\mu \leq \lambda \in \pi$ implies $\mu \in \pi$.

15.41 Corollary. *Denote by $\mathcal{F}(\Delta)$ (respectively $\mathcal{F}(\nabla)$) the categories of $\underline{\mathrm{gl}}(A)$-comodules that have a Δ-filtration (respectively ∇-filtration). Then:*

(1) *$\mathcal{F}(\Delta)$ and $\mathcal{F}(\nabla)$ are closed under tensor products.*
(2) *$M(\lambda) \in \mathcal{F}(\Delta) \cap \mathcal{F}(\nabla)$.*

So we see that the $M(\lambda)$ indeed play a role analogous to the tensor products of the $\bigwedge^i V$ for $\mathrm{GL}(V)$. In fact, it turns out that the linearization of $\mathcal{U}$ is equivalent to $\mathcal{V}$, and we even obtain an analogue of Theorem 15.27.

15.42 Theorem. *The monoidal functor*

$$M : \mathbb{K}\mathcal{U} \to \mathrm{rep}_{\mathbb{K}}\big(\underline{\mathrm{gl}}(A)\big) : \lambda \mapsto M(\lambda)$$

is fully faithful and its essential image is $\mathcal{V}$. Moreover, the derived version of M

$$M : \ \mathrm{Perf}(\mathcal{U}^{\mathrm{op}}) \to \mathbf{D}^b\left(\mathrm{rep}_{\mathbb{K}}\big(\underline{\mathrm{gl}}(A)\big)\right) \tag{15.27}$$

induced by $\mathbb{K}\mathcal{U}(-, \lambda) \mapsto M(\lambda)$ is an equivalence of monoidal triangulated categories.

15.43 Corollary. *The representation ring of $\underline{\mathrm{gl}}(A)$ is given by*

$$\mathbb{Z}\langle r_1, \ldots, r_{d-1}, r_d, r_d^{-1}\rangle$$

where r_i corresponds to $[R_i]$.

Proof. In a quasi-hereditary coalgebra, the costandard representations $[\nabla(\lambda)]$ form a $\mathbb{Z}$-basis of the representation ring $G_0\big(\underline{\mathrm{gl}}(A)\big)$. By Corollary 15.41(2), $M(\lambda) \in \mathcal{F}(\nabla)$, and one can show that they are related to the costandard representations by a unitriangular matrix. Hence, the $[M(\lambda)]$ also form a $\mathbb{Z}$-basis of $G_0\big(\underline{\mathrm{gl}}(A)\big)$. Since M is monoidal and maps r_i to $M(r_i) = R_i$, we are done. □

This gives yet more motivation for thinking of $\underline{\mathrm{gl}}(A)$ as a noncommutative version of the coordinate ring of $\mathrm{GL}(V)$, since the representation ring of GL_d is of the form $\mathbb{Z}[r_1, \ldots, r_{d-1}, r_d, r_d^{-1}]$.

15.5.3 *Co-Morita Equivalences*

Two bialgebras are said to be *co-Morita equivalent* if their monoidal categories of comodules are equivalent, as monoidal $\mathbb{K}$-linear categories. In Theorem 15.42, the domain category does not depend on the specific A we started with, but only on its global dimension d. In fact, the equivalence (15.27) may be used to transfer the standard t-structure on the derived category $\mathbf{D}^b\left(\mathrm{rep}_{\mathbb{K}}(\underline{\mathrm{gl}}(A))\right)$ to one on $\mathrm{Perf}(\mathcal{U}^{\mathrm{op}})$. This can be used to give an intrinsic description of the induced t-structure referring solely to properties of $\mathcal{U}$. As a corollary we obtain:

15.44 Theorem. *The category* $\mathrm{rep}_{\mathbb{K}}(\underline{\mathrm{gl}}(A))$ *as a monoidal category only depends on the global dimension of A and not on A itself.*

In other words by letting A vary we obtain nontrivial examples of co-Morita equivalent Hopf algebras [57, Section 5].

This somewhat curious corollary to Theorem 15.42 turns out to be a special case of a much more general phenomenon. Namely, given two monoidal functors $F, G\colon \mathcal{C} \to \mathrm{vect}_{\mathbb{K}}$ on a fixed rigid monoidal category $\mathcal{C}$, one can form an algebra $\mathrm{cohom}(F, G)$, by using a slight variation of Remark 15.3.

15.45 Theorem ([53]). *If* $\mathrm{cohom}(F, G) \neq 0$, *then cotensoring with it induces an equivalence*

$$\mathrm{rep}_{\mathbb{K}}(\mathrm{coend}(G)) \to \mathrm{rep}_{\mathbb{K}}(\mathrm{coend}(F))$$

of monoidal categories.

15.46 *Remark.* The relation $\mathrm{cohom}(F, G) \neq 0$ is actually an equivalence relation and divides up the monoidal functors $\mathcal{C} \to \mathrm{vect}_{\mathbb{K}}$ into connected components whose members yield co-Morita equivalent Hopf algebras.

Proof of Theorem 15.44 *using Theorem* 15.45. Consider two Koszul AS-regular algebras A and B, both of global dimension d. Section 15.5.2 provides us with two monoidal functors $M_A, M_B\colon \mathcal{U} \to \mathrm{vect}_{\mathbb{K}}$, such that

$$\mathrm{coend}(M_A) \cong \underline{\mathrm{gl}}(A),$$
$$\mathrm{coend}(M_B) \cong \underline{\mathrm{gl}}(B).$$

By Theorem 15.45, it suffices to show that $\mathrm{cohom}_{\mathcal{U}}(M_A, M_B) \neq 0$. Consider the (saturated) subset $\{1\} \subset A$, and the corresponding one-object category $\mathbf{1} \subset \mathcal{U}$. By a suitable analogue of Theorem 15.40 one shows that

$$\mathrm{cohom}_{\mathbf{1}}(M_A|_{\mathbf{1}}, M_B|_{\mathbf{1}}) \hookrightarrow \mathrm{cohom}_{\mathcal{U}}(M_A, M_B)\ .$$

Since $\mathbb{K} \cong \mathrm{cohom}_{\mathbf{1}}(M_A|_{\mathbf{1}}, M_B|_{\mathbf{1}}) \neq 0$, we are done. $\square$

15.6 Representations of $\underline{\mathrm{end}}(A)$

One might wonder whether the relation between $\underline{\mathrm{end}}(A)$ and $\underline{\mathrm{gl}}(A)$ is as close as the relation between $\mathcal{O}(M_n)$ and $\mathcal{O}(\mathrm{GL}(V))$ (introduced in Section 15.4.1). That this is indeed the case follows from the following proposition. Denote

$$\Lambda^+ := \langle r_1, \ldots, r_d \rangle \subset \Lambda .$$

15.47 Proposition. *The bialgebra $\underline{\mathrm{end}}(A)$ is the minimal subcoalgebra of $\underline{\mathrm{gl}}(A)$ whose representations have simple composition factors belonging to the set $\{L(\lambda)_{\lambda\in\Lambda^+}\}$.*

Proposition 15.47 allows us to recover most of the results from [42], which describe the representation theory of $\underline{\mathrm{end}}(A)$, for A Koszul. For the sake of exposition, we will restate their main result of [42] in the language of Section 15.5, and in the case when A is also AS-regular. To do this, we need to consider the monoidal subcategory $\mathcal{U}_\uparrow^+$ of $\mathcal{U}_\uparrow$, generated by the objects $r_1, \ldots, r_d$.

15.48 Theorem ([42, Theorem 4.3, Cor. 4.4]). *For a Koszul AS-regular algebra A of global dimension d, there is a monoidal functor*

$$M^+ \colon \mathrm{mod}\left(\mathcal{U}_\uparrow^{+,\mathrm{op}}\right) \to \mathrm{rep}_{\mathbb{K}}\left(\underline{\mathrm{end}}(A)\right) \colon \mathcal{U}_\uparrow^+(-,u) \mapsto M^+(u)$$

which is an equivalence of monoidal categories.

In fact, our techniques ensure that $\underline{\mathrm{end}}(A)$ is quasi-hereditary with $\Delta(\lambda) = M(\lambda)$. Moreover, the $\left(\Delta(\lambda)\right)_{\lambda\in\Lambda^+}$ form a system of projective generators for $\mathrm{rep}_{\mathbb{K}}\left(\underline{\mathrm{end}}(A)\right)$ and we obtain the equivalence.

The representation theory of $\underline{\mathrm{end}}(A)$ can be understood in terms of quivers with relations as follows. For every number $n \geq 0$, set $C_n = \mathbb{K}Q_n/I$ to be the quiver with relations corresponding to an n-dimensional (directed) hypercube Q_n with commuting faces. Depending on the global dimension d, we need to consider certain full subalgebras $C_{n,d} \subset C_n$, obtained by deleting some of the vertices of the hypercubes, and then there is an equivalence of (abelian) categories

$$\mathrm{rep}_{\mathbb{K}}\left(\underline{\mathrm{end}}(A)\right) \cong \mathrm{mod}\left(\mathcal{U}_\uparrow^{+,\mathrm{op}}\right) \cong \bigoplus_n \mathrm{mod}(C_{n,d}).$$

Rather than spelling out the somewhat contrived (though easy to implement in practice) combinatorial rule for constructing $C_{n,d}$, we refer the reader to [42].

15.7 Representations of $\underline{\mathrm{gl}}(\mathbb{K}[x, y])$

Theorem 15.44 tells us that to understand the representation theory of $\underline{\mathrm{gl}}(A)$, with A of global dimension d, it suffices to study the representations of $\underline{\mathrm{gl}}(\mathrm{Sym}_{\mathbb{K}}(V))$, with $\dim_{\mathbb{K}}(V) = d$, and the functor

$$\mathrm{rep}_{\mathbb{K}}\left(\underline{\mathrm{gl}}(\mathrm{Sym}_{\mathbb{K}}(V))\right) \to \mathrm{rep}_{\mathbb{K}}(\underline{\mathrm{gl}}(A))$$

that realizes the equivalence of monoidal categories.

Since $\mathrm{Sym}_{\mathbb{K}}(V)$ is a commutative ring, one might hope that there is an even closer connection between representations of $\underline{\mathrm{gl}}(\mathrm{Sym}_{\mathbb{K}}(V))$ and $\mathrm{rep}_{\mathbb{K}}(\mathrm{GL}(V))$. For this reason we will denote $\underline{\mathrm{gl}}(\mathrm{Sym}_{\mathbb{K}}(V))$ by $\mathcal{O}_{\mathrm{nc}}(\mathrm{GL}_d)$, which we think of as some sort of noncommutative coordinate ring of $\mathrm{GL}(V)$.

For V of dimension 1, we find

$$\mathcal{O}_{\mathrm{nc}}(\mathrm{GL}_1) = \mathcal{O}(T) = K[t, t^{-1}] ,$$

the coordinate ring of a one-dimensional torus, so $d = 2$ is the first interesting case. In this section we review the results from [55], where this example is treated in detail.

15.7.1 *A Noncommutative Version of the Borel and Torus Subgroups*

In Section 15.4.1, we saw the importance of the torus and Borel subgroups in the representation theory of $\mathrm{GL}(V)$. For $\mathcal{O}_{\mathrm{nc}}(\mathrm{GL}_2)$ it is possible to define analogues of the coordinate rings of these subgroups, using the explicit presentation from Example 15.13:

$$\begin{aligned} \mathcal{O}_{\mathrm{nc}}(B) &= \mathcal{O}_{\mathrm{nc}}(\mathrm{GL}_2)/(b) \cong \mathbb{K}\langle c, d^{\pm 1}\rangle[a^{\pm 1}] , \\ \mathcal{O}(T) &= \mathcal{O}_{\mathrm{nc}}(\mathrm{GL}_2)/(b, c) \cong \mathbb{K}[a^{\pm 1}, d^{\pm 1}] . \end{aligned}$$

Here $\mathcal{O}(T)$ is the (commutative) coordinate ring of a two-dimensional torus T. We identify its character group $X(T)$ with the Laurent monomials in a, d. By sending $r_2 \in \Lambda$ to $ad \in X(T)$ and $r_1 \in \Lambda$ to $d \in X(T)$ we obtain a map of monoids $\mathtt{wt}\colon \Lambda \to X(T)$. We can now easily imitate the construction of the induced representations.

If $t \in X(T)$ then there is an associated one-dimensional $\mathcal{O}(T)$-representation $\mathbb{K}_t$ which may also be viewed as a $\mathcal{O}_{\mathrm{nc}}(B)$-representation. Denote by $\mathrm{ind}_B^{\mathrm{GL}_2}$ the right adjoint to the restriction functor

$$\mathrm{Res}_B^{\mathrm{GL}_2}\colon \mathrm{rep}_{\mathbb{K}}\big(\mathcal{O}_{\mathrm{nc}}(\mathrm{GL}_2)\big) \to \mathrm{rep}_{\mathbb{K}}\big(\mathcal{O}_{\mathrm{nc}}(B)\big) .$$

Then we have the following result:

15.49 Theorem. *There is a decomposition*

$$\mathrm{ind}_B^{\mathrm{GL}_2}(\mathbb{K}_t) = \bigoplus_{\substack{\lambda\in\Lambda \\ \mathtt{wt}(\lambda)=t}} \nabla(\lambda) .$$

In particular, we see that $\mathrm{ind}_B^{\mathrm{GL}_2}(\mathbb{K}_t) = 0$ if $t \notin X(T)^+ := \mathrm{im}\,\mathtt{wt}$. This agrees with the commutative case where only dominant weights yield nonzero representations under induction. But we also see that in contrast to the commutative case here the induced representations are not indecomposable. However, they still yield all costandard comodules.

15.7.2 *The Simple Representations*

From the fact that $\mathcal{O}_{\mathrm{nc}}(\mathrm{GL}_2)$ is quasi-hereditary it follows by general theory that the simple comodules are of the form $L(\lambda) = \mathrm{im}\big(\Delta(\lambda) \to \nabla(\lambda)\big)$; this, in principle, reduces their study to a linear algebra problem.

This problem is usually difficult to solve, but in this particular case it is possible. The bialgebra $\underline{\mathrm{end}}(A)$ is the subalgebra (by Proposition 15.47) of $\mathcal{O}_{\mathrm{nc}}(\mathrm{GL}_2)$ generated by a, b, c, d, and we have:

15.50 Theorem. *Assume that* $\mathrm{char}(\mathbb{K}) = 0$. *All simple $\mathcal{O}_{\mathrm{nc}}(\mathrm{GL}_2)$-representations are repeated tensor products of simple $\underline{\mathrm{end}}(k[x, y])$-representations and their duals.*

The simple $\underline{\mathrm{end}}(\mathbb{K}[x,y])$-representations can be understood using Theorem 15.48, and were considered in [42]. They are tensor products of $\big(\mathrm{Sym}^n(V)\big)_{n\in\mathbb{N}}$ and $\bigwedge^2 V$, where V denotes the standard representation. Thus every simple $\mathcal{O}_{\mathrm{nc}}(\mathrm{GL}_2)$-representation is a tensor product of these basic representations and their duals. It is somewhat intricate to characterize which among those tensor products are simple, but this is achieved in [55].

15.8 More Examples of Universal Hopf Algebras

Various other types of universal Hopf algebras have also been considered in the literature, see for example [9, 11, 12, 15, 49, 67]. Most of them arise as quotients of $\underline{\mathrm{gl}}(A)$, for some algebra A. In this section we will discuss one such example and we will also give some more comments on the commutative case which was touched upon in Theorem 15.27.

15.8.1 *The Universal Quantum Group of a Nondegenerate Bilinear Form*

In [22], Dubois-Violette and Launer introduced the *universal quantum group of a nondegenerate bilinear form*. Their definition is equivalent to the following.

15.51 Definition. Given a vector space V of $1 < \dim(V) < \infty$, and a nondegenerate bilinear form $b\colon V \otimes V \to \mathbb{K}$, the universal quantum group $H(b)$ of b is the universal Hopf algebra coacting on V making b into an $H(b)$-comodule morphism (for the trivial comodule structure on $\mathbb{K}$).

15.52 *Example.* For $q \in \mathbb{K}^*$, consider the bilinear form given by

$$b = \begin{pmatrix} 0 & 1 \\ -q^{-1} & 0 \end{pmatrix} .$$

One then computes that $H(b) = \mathcal{O}_q(\mathrm{SL}_2)$, the quantum coordinate ring of SL_2 (see for example, [39]).

Setting $A = TV^*/(b)$, we obtain a Koszul AS-regular algebra, see [70, Proposition 1.1], and one verifies that there is an isomorphism of Hopf algebras

$$\underline{\mathrm{sl}}(A) \cong H(b) , \tag{15.28}$$

where $\underline{\mathrm{sl}}(A)$ was introduced in Section 9.5.

There is a close connection between $H(b)$ and the famous *Temperley–Lieb category*. The Temperley–Lieb category is best known for its attractive graphical model, based on planar nonintersecting strands (see [2]), but abstractly it can be characterized by the following presentation

$$\begin{aligned} \mathcal{U} = \langle v \,|\, 1 \xrightarrow{\phi} vv, vv \xrightarrow{\psi} 1 \,|\, v \xrightarrow{1\otimes\phi} vvv \xrightarrow{\psi\otimes 1} v = \mathrm{id}_v = v \xrightarrow{\phi\otimes 1} vvv \xrightarrow{1\otimes\psi} v, \\ v \xrightarrow{1\otimes\phi} vvv \xrightarrow{1\otimes\psi} v = v \xrightarrow{\phi\otimes 1} vvv \xrightarrow{\psi\otimes 1} v\rangle_\otimes . \end{aligned} \tag{15.29}$$

The generators ψ and ϕ correspond to cups and caps in the graphical model. The third relation ensures that the "circle" $\eta := \psi\phi \in \mathrm{End}(1)$ acts centrally on $\mathcal{U}$. One has

$$H(b) = \mathrm{coend}(F_b), \tag{15.30}$$

where $F_b\colon \mathcal{U} \to \mathrm{vect}_k$ is the monoidal functor with $F_b(v) = V$, $F_b(\psi) = b$ and $F_b(\phi)\colon \mathbb{K} \to V \otimes V$ being dual to b.

The image $q(b) := F_b(\eta) \in \mathrm{End}(\mathbb{K}) = \mathbb{K}$ of η under F_b is called the *quantum dimension* of V (it can be zero!). It divides up the space of monoidal functors $\mathcal{U} \to \mathrm{vect}_{\mathbb{K}}$ into connected components.

15.53 Theorem ([10, Theorem 1.1]). *Let b, b' be two nondegenerate bilinear forms. Then there is a monoidal equivalence*

$$\mathrm{rep}_{\mathbb{K}}\big(H(b)\big) \cong \mathrm{rep}_{\mathbb{K}}\big(H(b')\big)$$

if and only if $q(b) = q(b')$. In particular, if we choose $q \in \mathbb{K}^*$ so that $q + q^{-1} = q(b)$, then there is an equivalence of monoidal categories

$$\mathrm{rep}_{\mathbb{K}}\big(H(b)\big) \cong \mathrm{rep}_{\mathbb{K}}\big(\mathcal{O}_q(\mathrm{SL}_2)\big)\ .$$

Theorem 15.53 is in stark contrast with Theorem 15.44, despite the fact that (15.28) seems to indicate that $\underline{\mathrm{gl}}(A)$ and $H(b)$ are quite close.

15.8.2 *Spiders and Representations of* GL(V)

We briefly indicate how the techniques from Section 15.5 fit in well with the planar diagrammatic approach to the representation theory of algebraic groups and Lie algebras.

Recall that in Theorem 15.27 we showed that

$$\mathcal{V} = \langle \textstyle\bigwedge^i V \mid i = 1, \dots, n \rangle_{\otimes} \subset \mathrm{rep}_{\mathbb{K}}(G)\ ,$$

for $G = \mathrm{GL}(V)$ and $\dim(V) = n$, gives rise to a derived equivalence

$$F\colon \mathrm{Perf}(\mathcal{V}^{\mathrm{op}}) \to \mathbf{D}^b\big(\mathrm{rep}_{\mathbb{K}}(G)\big)\colon \mathcal{V}(-, u) \mapsto F(u)\ .$$

Motivated by Theorem 15.42, one might wonder if the category $\mathcal{V}$ has a nice combinatorial presentation as monoidal category.

We will assume that $\mathrm{char}(\mathbb{K}) = 0$,[8] though see Remark 15.54. As generating morphisms one can take the natural maps

$$\Gamma_{i,j} : \bigwedge^i V \otimes \bigwedge^j V \to \bigwedge^{i+j} V,$$
$$\Sigma_{i,j} : \bigwedge^{i+j} V \to \bigwedge^i V \otimes \bigwedge^j V.$$

The relations among the $\Gamma_{i,j}$ and $\Sigma_{i,j}$ can be determined using *skew Howe duality*, allowing for a planar diagrammatic description of the category $\mathcal{V}$. A morphism between tensor products of the $\bigwedge^i V$ is encoded as a certain kind of oriented graph, resulting in a category called the $\mathrm{GL}(V)$-*spider*. See [13, Theorem 3.3.1] for more details on skew Howe duality and spiders.

15.54 *Remark.* For $\mathrm{char}(\mathbb{K}) > 0$, there is a version of skew Howe duality by Adamovich and Rybnikov [3], but the resulting algebraic presentation is quite involved. The diagrammatic approach seems more flexible, see [24]. R. Howe formulated what is now called "Howe duality" from a unifying point of view of Lie superalgebras. For a lucid exposition of Howe duality and its further development, see [43].

[8] In this case there is no need for derived categories: closing $\mathcal{V}$ under direct summands recovers $\mathrm{rep}_{\mathbb{K}}(G)$ since representations of G are completely reducible.

Bibliography

1. Abe, E.: Hopf Algebras, *Cambridge Tracts in Math.*, vol. 74. Cambridge Univ. Press, Cambridge–New York (1980)
2. Abramsky, S.: Temperley–Lieb algebra: From knot theory to logic and computation via quantum mechanics. In: G. Chen, L. Kauffman, S.J. Lomonaco (eds.) Mathematics of Quantum Computation and Quantum Technology, Chapman & Hall/CRC Appl. Math. Nonlinear Sci. Ser., pp. 515–558. CRC Press, Boca Raton, FL (2008)
3. Adamovich, A.M., Rybnikov, G.L.: Tilting modules for classical groups and Howe duality in positive characteristic. Transform. Groups **1**(1-2), 1–34 (1996). https://doi.org/10.1007/BF02587733
4. Backelin, J.: On the rates of growth of the homologies of Veronese subrings. In: J.-E. Roos (ed.) Algebra, Algebraic Topology and Their Interactions (Stockholm, 1983), *Lecture Notes in Math.*, vol. 1183, pp. 79–100. Springer, Berlin (1986). https://doi.org/10.1007/BFb0075451
5. Backelin, J., Fröberg, R.: Koszul algebras, Veronese subrings and rings with linear resolutions. Rev. Roumaine Math. Pures Appl. **30**(2), 85–97 (1985)
6. Barr, M.: $*$-Autonomous Categories, *Lecture Notes in Math.*, vol. 752. Springer, Berlin (1979). With an appendix by P.H. Chu.
7. Van den Bergh, M.: Calabi-Yau algebras and superpotentials. Selecta Math. (N.S.) **21**(2), 555–603 (2015). https://doi.org/10.1007/s00029-014-0166-6
8. Bernshtein, I.N., Gelfand, I.M., Gelfand, S.I.: Structure of representations that are generated by vectors of highest weight (in Russian). Funckcional. Anal. i Priložen. **5**(1), 1–9 (1971). English transl. Functional Anal. Appl. **5**, 1–8 (1971)
9. Bichon, J.: Cosovereign Hopf algebras. J. Pure Appl. Algebra **157**(2-3), 121–133 (2001). https://doi.org/10.1016/S0022-4049(00)00024-4
10. Bichon, J.: The representation category of the quantum group of a non-degenerate bilinear form. Comm. Algebra **31**(10), 4831–4851 (2003). https://doi.org/10.1081/AGB-120023135
11. Bichon, J.: Co-representation theory of universal co-sovereign Hopf algebras. J. Lond. Math. Soc. (2) **75**(1), 83–98 (2007). https://doi.org/10.1112/jlms/jdl007

Y. I. Manin, *Quantum Groups and Noncommutative Geometry*,
CRM Short Courses, https://doi.org/10.1007/978-3-319-97987-8

12. Bichon, J., Dubois-Violette, M.: The quantum group of a preregular multilinear form. Lett. Math. Phys. **103**(4), 455–468 (2013)
13. Cautis, S., Kamnitzer, J., Morrison, S.: Webs and quantum skew Howe duality. Math. Ann. **360**(1-2), 351–390 (2014). https://doi.org/10.1007/s00208-013-0984-4
14. Chari, V., Pressley, A.: A Guide to Quantum Groups. Cambridge Univ. Press, Cambridge (1995)
15. Chirvasitu, A., Walton, C., Wang, X.: On quantum groups associated to a pair of preregular forms. arXiv:1605.06428
16. Deligne, P., Milne, J.S.: Hodge cycles, motives, and Shimura varieties. In: Hodge Cycles, Motives, and Shimura Varieties [17], chap. 2, pp. 101–228
17. Deligne, P., Milne, J.S., Ogus, A., Shih, K.-y.: Hodge Cycles, Motives, and Shimura Varieties, *Lecture Notes in Math.*, vol. 900. Springer, Berlin–New York (1982)
18. Dlab, V., Ringel, C.M.: The module theoretical approach to quasi-hereditary algebras. In: Representations of Algebras and Related Topics (Kyoto, 1990), *London Math. Soc. Lecture Note Ser.*, vol. 168, pp. 200–224. Cambridge Univ. Press, Cambridge (1992)
19. Doebner, H.D., Hennig, J.D., Lücke, W.: Mathematical guide to quantum groups. In: Quantum Groups (Clausthal, 1989), *Lecture Notes in Phys.*, vol. 370, pp. 29–63. Springer, Berlin (1990). https://doi.org/10.1007/3-540-53503-9_40
20. Drinfeld, V.G.: On quadratic commutation relations in the quasiclassical case (in Russian). In: Mathematical Physics, Functional Analysis, pp. 25–34, 143. "Naukova Dumka", Kiev (1986)
21. Drinfeld, V.G.: Quantum groups. In: Proceedings of the International Congress of Mathematicians, Vol. 1, 2 (Berkeley, CA, 1986), pp. 798–820. Amer. Math. Soc., Providence, RI (1987)
22. Dubois-Violette, M., Launer, G.: The quantum group of a nondegenerate bilinear form. Phys. Lett. B **245**(2), 175–177 (1990). https://doi.org/10.1016/0370-2693(90)90129-T
23. Egorov, G.: How to superize $\mathfrak{gl}(\infty)$. In: Topological and Geometrical Methods in Field Theory (Turku, 1991), pp. 135–146. World Sci. Publ., Singapore (1992)
24. Elias, B.: Light ladders and clasp conjectures. arXiv:1510.06840
25. Etingof, P., Gelaki, S.: Isocategorical groups. Internat. Math. Res. Notices **2001**(2), 59–76 (2001). https://doi.org/10.1155/S1073792801000046
26. Faddeev, L.D.: Integrable models in $(1 + 1)$-dimensional quantum field theory. In: Recent Advances in Field Theory and Statistical Mechanics (Les Houches, 1982), pp. 561–608. North-Holland, Amsterdam (1984)
27. Faddeev, L.D., Takhtajan, L.A.: Hamiltonian Methods in the Theory of Solitons. Springer Ser. Soviet Math. Springer, Berlin (1987). https://doi.org/10.1007/978-3-540-69969-9
28. Gabriel, P.: Des catégories abéliennes. Bull. Soc. Math. France **90**, 323–448 (1962)
29. Grozman, P., Leites, D., Shchepochkina, I.: Lie superalgebras of string theories. Acta Math. Vietnam. **26**(1), 27–63 (2001). arXiv:hep-th/970212
30. Gurevich, D.I.: Algebraic aspects of the quantum Yang–Baxter equation. Algebra i Analiz **2**(4), 119–148 (1990)
31. Gurevich, D.I., Pyatov, P.N., Saponov, P.A.: Hecke symmetries and characteristic relations on reflection equation algebras. Lett. Math. Phys. **41**(3), 255–264 (1997). https://doi.org/10.1023/A:1007386006326

32. Gurevich, D.I., Saponov, P.A.: Braided affine geometry and q-analogs of wave operators. J. Phys. A **42**(31), 313,001, 51 pp. (2009). https://doi.org/10.1088/1751-8113/42/31/313001. arXiv:0906.1057
33. He, J.W., Torrecillas, B., Van Oystaeyen, F., Zhang, Y.: Dualizing complexes of Noetherian complete algebras via coalgebras. Comm. Algebra **42**(1), 271–285 (2014). https://doi.org/10.1080/00927872.2012.713063
34. Humphreys, J.E.: Representations of Semisimple Lie Algebras in the BGG Category $\mathcal{O}$, *Grad. Stud. Math.*, vol. 94. Amer. Math. Soc., Providence, RI (2008). https://doi.org/10.1090/gsm/094
35. Izumi, M., Kosaki, H.: On a subfactor analogue of the second cohomology. Rev. Math. Phys. **14**(7-8), 733–757 (2002). https://doi.org/10.1142/S0129055X02001375
36. Jimbo, M.: A q-difference analogue of $U(\mathfrak{g})$ and the Yang–Baxter equation. Lett. Math. Phys. **10**(1), 63–69 (1985). https://doi.org/10.1007/BF00704588
37. Jimbo, M.: A q-analogue of $U(\mathfrak{gl}(N+1))$, Hecke algebra, and the Yang–Baxter equation. Lett. Math. Phys. **11**(3), 247–252 (1986). https://doi.org/10.1007/BF00400222
38. Kac, V.G.: Classification of supersymmetries. In: Proceedings of the International Congress of Mathematicians, Vol. I (Beijing, 2002), pp. 319–344. Higher Ed. Press, Beijing (2002)
39. Kassel, Ch.: Quantum Groups, *Grad. Texts in Math.*, vol. 155. Springer, New York (1995). https://doi.org/10.1007/978-1-4612-0783-2
40. Kassel, Ch., Rosso, M., Turaev, V.: Quantum Groups and Knot Invariants, *Panor. Synthèses*, vol. 5. Soc. Math. France, Paris (1997)
41. Krause, H.: The spectrum of a module category. Mem. Amer. Math. Soc. **149**(707) (2001). https://doi.org/10.1090/memo/0707
42. Kriegk, B., Van den Bergh, M.: Representations of non-commutative quantum groups. Proc. Lond. Math. Soc. (3) **110**(1), 57–82 (2015). https://doi.org/10.1112/plms/pdu042
43. Leites, D., Shchepochkina, I.: The Howe duality and Lie superalgebras. In: S. Duplij, J. Wess (eds.) Noncommutative Structures in Mathematics and Physics (Kiev, 2000), *NATO Sci. Ser. II Math. Phys. Chem.*, vol. 22, pp. 93–111. Kluwer, Dordrecht (2001). https://doi.org/10.1007/978-94-010-0836-5_8. arXiv:math/0202181
44. Löfwall, C.: On the subalgebra generated by the one-dimensional elements in the Yoneda Ext-algebra. In: Algebra, Algebraic Topology and Their Interactions (Stockholm, 1983), *Lecture Notes in Math.*, vol. 1183, pp. 291–338. Springer, Berlin (1986). https://doi.org/10.1007/BFb0075468
45. Lyubashenko, V.V.: Hopf algebras and vector-symmetries. Uspekhi Mat. Nauk **41**(5(251)), 185–186 (1986). English transl., Russian Math. Surveys **41**(5), 153–154 (1986)
46. Manin, Yu.I.: Some remarks on Koszul algebras and quantum groups. Ann. Inst. Fourier (Grenoble) **37**(4), 191–205 (1987)
47. Manin, Yu.I.: Gauge Field Theory and Complex Geometry, *Grundlehren Math. Wiss.*, vol. 289. Springer, Berlin (1988). https://doi.org/10.1007/978-3-662-07386-5
48. Mathieu, O.: Filtrations de G-modules. C. R. Acad. Sci. Paris Sér. I Math. **309**(6), 273–276 (1989)
49. Mrozinski, C.: Quantum groups of GL(2) representation type. J. Noncommut. Geom. **8**(1), 107–140 (2014). https://doi.org/10.4171/JNCG/150

50. Neretin, Yu.A.: Gauss–Berezin integral operators, spinors over orthosymplectic supergroups, and Lagrangian super-Grassmannians. arXiv:0707.0570
51. Pareigis, B.: Endomorphism bialgebras of diagrams and of noncommutative algebras and spaces. In: J. Bergen, S. Montgomery (eds.) Advances in Hopf Algebras (Chicago, IL, 1992), *Lecture Notes in Pure and Appl. Math.*, vol. 158, pp. 153–186. Dekker, New York (1994)
52. Priddy, S.B.: Koszul resolutions. Trans. Amer. Math. Soc. **152**, 39–60 (1970). https://doi.org/10.2307/1995637
53. Raedschelders, Th.: Manin's universal Hopf algebras and highest weight categories. Ph.D. Thesis, Vrije Universiteit Brussel (2017)
54. Raedschelders, Th., Van den Bergh, M.: The Manin Hopf algebra of a Koszul Artin–Schelter regular algebra is quasi-hereditary. Adv. Math. **305**, 601–660 (2017). https://doi.org/10.1016/j.aim.2016.09.017
55. Raedschelders, Th., Van den Bergh, M.: The representation theory of non-commutative $\mathcal{O}(\mathrm{GL}_2)$. J. Noncommut. Geom. **11**(3), 845–885 (2017). https://doi.org/10.4171/JNCG/11-3-3
56. Reshetikhin, N.Yu., Takhtajan, L.A., Faddeev, L.D.: Quantization of Lie groups and Lie algebras. Algebra i Analiz **1**(1), 178–206 (1989)
57. Schauenburg, P.: Hopf bi-Galois extensions. Comm. Algebra **24**(12), 3797–3825 (1996). https://doi.org/10.1080/00927879608825788
58. Sklyanin, E.K.: Quantum variant of the method of the inverse scattering problem. Zap. Nauchn. Sem. Leningrad. Otdel. Mat. Inst. Steklov. (LOMI) **95**, 55–128 (1980)
59. Sklyanin, E.K., Takhtajan, L.A., Faddeev, L.D.: Quantum inverse problem method. I. Teoret. Mat. Fiz. **40**(2), 194–220 (1979)
60. Smith, S.P.: Some finite-dimensional algebras related to elliptic curves. In: R. Bautista, R. Martínez-Villa, J.A. de la Peña (eds.) Representation Theory of Algebras and Related Topics (Mexico City, 1994), *CMS Conf. Proc.*, vol. 19, pp. 315–348. Amer. Math. Soc., Providence, RI (1996)
61. Soibelman, Y.S.: Irreducible representations of the algebra of functions on the quantum group $\mathrm{SU}(n)$ and Schubert cells. Dokl. Akad. Nauk SSSR **307**(1), 41–45 (1989)
62. Stephenson, D.R., Zhang, J.J.: Growth of graded Noetherian rings. Proc. Amer. Math. Soc. **125**(6), 1593–1605 (1997). https://doi.org/10.1090/S0002-9939-97-03752-0
63. Takeuchi, M.: Morita theorems for categories of comodules. J. Fac. Sci. Univ. Tokyo Sect. IA Math. **24**(3), 629–644 (1977)
64. Vaksman, L.L., Soibelman, Y.S.: An algebra of functions on the quantum group $\mathrm{SU}(2)$. Funktsional. Anal. i Prilozhen. **22**(3), 1–14, 96 (1988). https://doi.org/10.1007/BF01077623
65. Verevkin, A.B.: Serre cohomology of noncommutative rings and Morita equivalence. Uspekhi Mat. Nauk **44**(5(269)), 157–158 (1989). https://doi.org/10.1070/RM1989v044n05ABEH002285
66. Verevkin, A.B.: On a noncommutative analogue of the category of coherent sheaves on a projective scheme. In: Algebra and Analysis (Tomsk, 1989), *Amer. Math. Soc. Transl. Ser. 2*, vol. 151, pp. 41–53. Amer. Math. Soc., Providence, RI (1992). https://doi.org/10.1090/trans2/151/02

67. Walton, C., Wang, X.: On quantum groups associated to non-Noetherian regular algebras of dimension 2. Math. Z. **284**(1-2), 543–574 (2016). https://doi.org/10.1007/s00209-016-1666-1
68. Woronowicz, S.L.: Compact matrix pseudogroups. Comm. Math. Phys. **111**(4), 613–665 (1987)
69. Woronowicz, S.L.: Twisted SU(2) group. An example of a noncommutative differential calculus. Publ. Res. Inst. Math. Sci. **23**(1), 117–181 (1987). https://doi.org/10.2977/prims/1195176848
70. Zhang, J.J.: Non-Noetherian regular rings of dimension 2. Proc. Amer. Math. Soc. **126**(2), 1645–1653 (1998). https://doi.org/10.1090/S0002-9939-98-04480-3

Index

Y. I. Manin, *Quantum Groups and Noncommutative Geometry*,
CRM Short Courses, https://doi.org/10.1007/978-3-319-97987-8

Printed in the United States
By Bookmasters